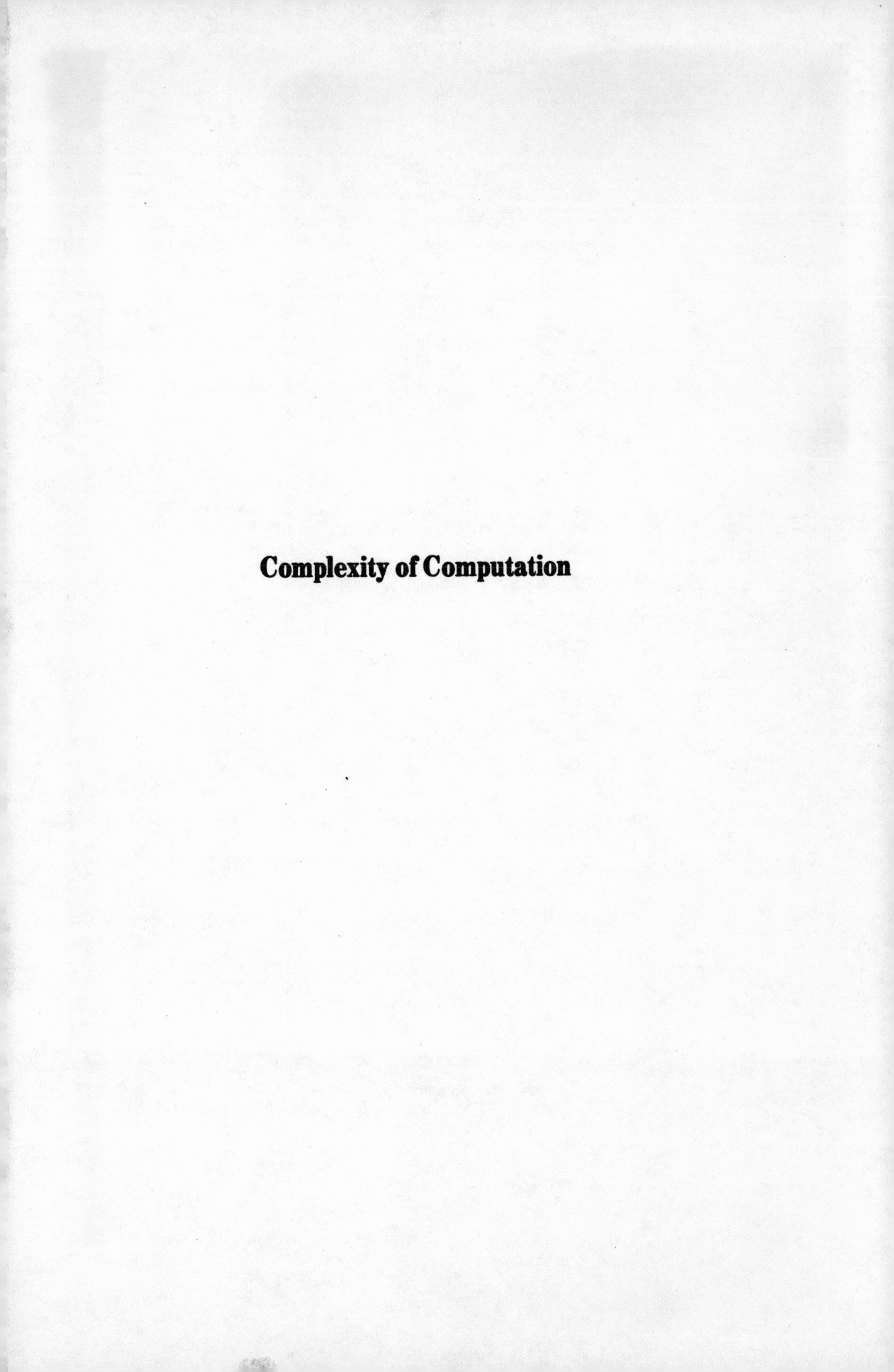

Complexity of Computation

VOLUME VII

SIAM-AMS PROCEEDINGS

Complexity of Computation

AMERICAN MATHEMATICAL SOCIETY
PROVIDENCE, RHODE ISLAND
1974

PROCEEDINGS OF A SYMPOSIUM IN APPLIED MATHEMATICS
OF THE AMERICAN MATHEMATICAL SOCIETY AND
THE SOCIETY FOR INDUSTRIAL AND APPLIED MATHEMATICS

Held in New York City
April 18–19, 1973

Edited by
Richard M. Karp

Prepared by the American Mathematical Society
with partial support of National Science Foundation grant GP–38244
and Office of Naval Research contract NONR(G) 00016–73

Library of Congress Cataloging in Publication Data CIP

Main entry under title:

Complexity of computation.

(SIAM–AMS proceedings; v. 7)

"Proceedings of a symposium in applied mathematics of the American Mathematical Society and the Society for Industrial and Applied Mathematics held in New York City April 18–19, 1973."

Includes bibliographies.

1. Machine theory–Congresses. 2. Electronic data processing–Mathematics–Congresses. 3. Electronic data processing–Numerical analysis. I. Karp, Richard M., 1935– ed. II. American Mathematical Society. III. Society for Industrial and Applied Mathematics. IV. Series: Society for Industrial and Applied Mathematics. SIAM–AMS proceedings; v. 7.

QA267.C58 519.4 74-22062
ISBN 0-8218-1327-7

AMS (MOS) subject classifications (1970). Primary 68A10, 68A20;
Secondary 02E10, 02F15, 68A25.

Contents

Preface

During the last decade computational complexity has become one of the most active research areas within the mathematical theory of computation. Workers in computational complexity seek to derive efficient algorithms for computational problems of practical interest, to prove the optimality of particular algorithms relative to well-defined measures of computational efficiency, and to derive general lower bounds on the time or space intrinsically necessary for the performance of computational tasks. The specific problems considered are drawn from diverse areas, including numerical computation, symbolic algebraic computation, combinatorics, computational logic and the manipulation of data structures. The mathematical tools called upon are correspondingly diverse, ranging from algebraic geometry through computability theory. Nevertheless, some characteristic proof techniques and approaches to algorithm construction are emerging as complexity theory matures and strives for unification.

Two of the papers in the present volume concern the relation between deterministic and nondeterministic computing devices. Hartmanis and Hunt discuss the so-called LBA problem—whether nondeterministic Turing machines operating within linear space have more power as recognizers than deterministic linear-space Turing machines. While the problem remains unsolved, this paper reduces it to determining whether certain specific decision problems can be solved deterministically in linear space; one such decision problem is the equivalence of regular expressions. The paper by Fagin defines generalized spectra and shows that they are essentially coextensive with languages recognized by nondeterministic Turing machines operating in polynomial time. Using a mixture of techniques from logic and automata theory, he exhibits specific "complete" generalized spectra. These results are motivated in part by the open $\mathcal{P}$ vs. $\mathcal{NP}$ question: whether nondeterministic Turing machines operating in polynomial time can recognize languages not recognizable by deterministic polynomial-time Turing machines.

Fischer and Rabin derive lower bounds on the computational complexity of the decision problem and on the inherent length of proofs for two classical decidable theories of logic: the first-order theory of the real numbers under addition, and Presburger arithmetic. Their results may be construed as implying that no decision procedure for either of these theories can be guaranteed to operate within a practical time bound, and no proof format can guarantee proofs of convenient length.

The papers by Aanderaa and by Paterson, Fischer and Meyer employ an "overlap" technique for deriving lower bounds on computing time. Aanderaa settles a long-standing conjecture by showing that, for real-time computation, k-tape Turing machines are more powerful than $(k - 1)$-tape Turing machines, for all $k \geqslant 2$. Paterson, Fischer and Meyer derive new lower bounds on the time required for the multiplication of numbers by on-line multi-tape Turing machines.

The paper by Fischer and Paterson exploits a formal similarity between pattern matching in strings and integer multiplication to derive a new, asymptotically efficient algorithm for a version of the former problem.

The last three papers in the volume study the computational complexity of fundamental processes in numerical computation. Gentleman shows that no method of expanding a determinant by minors requires fewer multiplications than expanding by column minors. Schultz studies the worst-case error bounds that can be achieved by certain linear approximations to continuous functions. Kung and Traub derive lower and upper bounds on the efficiency of iterative algorithms for approximating real numbers.

This volume is the proceedings of a symposium held in New York City on April 18 and 19, 1973, under the joint sponsorship of the American Mathematical Society and the Society for Industrial and Applied Mathematics. Financial support was provided by the National Science Foundation and the Office of Naval Research. Thanks are due to Stephen A. Cook, John E. Hopcroft and Shmuel Winograd for their help in the selection of the speakers at that symposium.

RICHARD M. KARP
EDITOR

SEPTEMBER 1974

SIAM-AMS PROCEEDINGS
Volume 7
1974

The LBA Problem And Its Importance In The Theory Of Computing

J. Hartmanis* and H. B. Hunt, III**

Abstract. In this paper we study the classic problem of determining whether the deterministic and nondeterministic context-sensitive languages are the same or, equivalently, whether the languages accepted by deterministic and nondeterministic linearly bounded automata are the same. We show that this problem is equivalent to several other natural problems in the theory of computing and that the techniques used on the LBA problem have several other applications in complexity theory. For example, we show that there exists a hardest tape recognizable nondeterministic context-sensitive language L_1 such that L_1 is a deterministic context-sensitive language if and only if the deterministic and nondeterministic context-sensitive languages are the same. We show furthermore that many decision problems about sets described by regular expressions are instances of these tape-hardest recognizable context-sensitive languages. Thus, it follows that nondeterminism in Turing machine computations (using at least linear tape) cannot save memory over deterministic Turing machine computations if and only if the equivalence of regular expressions can be decided by a deterministic linearly bounded automaton. It also follows that the equivalence of regular expressions can be decided by a nondeterministic linearly bounded automaton if and only if the family of context-sensitive languages is closed under complementation.

AMS (MOS) subject classifications (1970). Primary 68A20, 68A10; Secondary 02F15, 02F20.

*This research has been supported in part by the National Science Foundation grant GJ-33171X.

**This research was supported by a National Science Foundation Fellowship in Computer Science.

1. Introduction. In this section we sketch the history of the LBA problem and outline the results in this paper.

Linearly bounded automata were first defined and investigated by John Myhill in 1960 [13]. As Myhill points out, the definition of a linear bounded automaton was motivated by an observation made by Rabin and Scott about two-way finite automata with erasing. This remark appeared in a technical report on which the well-known Rabin and Scott paper *Finite automata and their decision problems* was based. The observation was that two-way finite automata, which can erase imput symbols, can accept nonregular sets and that the equivalence problem for these automata is recursively undecidable. These observations never appeared in the published paper, but the short paragraph in the original technical report sufficed to convince Myhill that this model with erasing only was artificial and that the automation should be permitted to erase and print on the tape squares occupied by the initial input word. Thus these automata are just one-tape Turing machines which can use for computation as much tape as is needed to write down the input word. Since this definition bounds the available tape linearly to the length of the input word Myhill called them linearly bounded automata.

The importance of linearly bounded automata was further emphasized when their connection with language theory was discovered. In the late fifties and early sixties Chomsky initiated an intensive study of formal languages and defined four classes of grammars with the corresponding languages: the regular, context-free, context-sensitive and recursively enumerable languages. After it was realized in 1962 that the context-free languages were exactly the languages accepted by nondeterministic push-down automata, the regular, context-free and recursively-enumerable languages could all be defined by their grammars or, equivalently, by the automata which accepted them. The context-sensitive languages remained the only exception.

In 1963 Landweber [10] showed that every set or language accepted by a deterministic linearly bounded automaton was a context-sensitive language. In 1964 Kuroda [9] introduced the nondeterministic linearly bounded automaton and showed that the family of languages accepted by the nondeterministic linearly bounded automata is exactly the same as the family of languages generated by the context-sensitive grammars.

These results revealed another natural connection between families of formal languages and families of automata; but they also raised the now classic LBA problem (or the first LBA problem):

> Are the languages accepted by deterministic and nondeterministic linearly bounded automata the same? Or equivalently, are the deterministic and nondeterministic context-sensitive languages the same? Abbreviated, DCSL = NDCSL?

If DCSL = NDCSL then the family of context-sensitive languages is closed under complementation. On the other hand, it still could happen that DCSL ≠ NDCSL but the family of context-sensitive languages is closed under complementation. Thus we are led to the second LBA problem:

> Are the context-sensitive languages closed under complementation?

Both of these problems are basically problems about the minimal amount of memory needed to perform a computation. In general, such problems are quite difficult and so far in computational complexity theory we have had little success in determining lower complexity bounds for specific computations. The above-mentioned LBA problems appear to be no exception. At the same time, our inability to answer them indicates that we have not yet understood the nature of nondeterministic computations.

Considerable progress on the first LBA problem was made in 1969 by W. Savitch in his doctoral dissertation [14]. Savitch showed that every nondeterministic Turing machine using $L(n)$-tape, $L(n) \geqslant \log n$, can be simulated by a deterministic Turing machine using no more than $[L(n)]^2$-tape. Thus the nondeterministic context-sensitive languages can all be recognized by n^2-tape bounded deterministic Turing machines. The result was suprising since all previous simulation methods required an exponential amount of tape. Furthermore, Savitch showed that there exists one nondeterministic $L(n) = (\log n)$-tape recognizable language L_0 such that if L_0 is recognizable deterministically in $(\log n)$-tape, then, for all tape bounds $L(n)$, $L(n) \geqslant \log n$, the nondeterministic and deterministic recognizable languages are the same. Thus if nondeterminism can be eliminated for the $(\log n)$-tape recognizable language L_0 then DCSL = NDCSL and we see that we have a sufficient condition for the LBA problem. Unfortunately, this was shown only to be a sufficient condition for DCSL = NDCSL.

In this paper we show that we can find necessary and sufficient conditions for DCSL = NDCSL in terms of one nondeterministic context-sensitive language by constructing a hardest deterministic tape recognizable context-sensitive language L_1. Thus we get that DCSL = NDCSL if and only if L_1 is a deterministic context-sensitive language.

Similarly, the family of context-sensitive languages is closed under complementation if and only if the complement of $L_1, \bar{L}_1$, is a nondeterministic context-sensitive language.

Actually the results are stronger in that DCSL = NDCSL implies that the deterministic and nondeterministic tape bounded computations are the same for all tape bounds $L(n) \geqslant n$. Furthermore, there exists a recursive translation which maps every nondeterministic Turing machine onto a deterministic one using no more tape than the nondeterministic one (provided it used at least linear tape).

Similarly, if the family of context-sensitive languages is closed under complementation, then there exists a recursive translation which maps every lba onto another lba accepting exactly those sequences not accepted by the first.

Next we show that the LBA problems can be reduced to equivalent problems about very simple nonwriting automata or flowchart computations. Consider finite automata with k read-only heads which can move in both directions on the input and sense when two heads are scanning the same tape square. Then, utilizing our previous results and an observation by Savitch, we show that there exists a language L_2 over a one-symbol alphabet, $L_2 \subseteq a^*$, which is recognizable by a 7-head nondeterministic finite automaton and has the property that L_2 is recognizable by a k-head deterministic finite automaton if and only if DCSL = NDCSL.

Again, $\bar{L}_2$ is recognizable by a k-head nondeterministic finite automaton if and only if the family of CSL's is closed under complementation which happens iff the family of languages over a one-symbol alphabet recognizable by multi-head automata is closed under complementation.

Thus we will show that if nondeterminism can be eliminated in one specific 7-head finite automaton by using more heads then it can be eliminated in all Turing machine computations using no less than linear tape. A similar result holds for flowcharts where we must eliminate nondeterminism by using more variables.

To relate the LBA problems to a different problem area we show that the complexity of the LBA problems is equivalent to many decision problems about sets described by regular expressions. In this case the proofs exploit an observation due to Meyer and Stockmeyer [12] about the descriptive power of regular expressions. It turns out that for any nondeterministic lba M_i there exists a deterministic lba which for any input y to M_i can write down a regular expression $R(y)$ describing the set of all invalid computations of M_i on input y. Therefore the input y is accepted by M_i if and only if there is a valid computation by M_i on y, which happens if and only if $L[R(y)] \neq \Sigma^*$, where $L(R)$ denotes the language described by R. Thus we see that if a deterministic lba can check whether a regular expression describes a set not equal to Σ^*, every nondeterministic lba M_i can be replaced by a deterministic lba, using the above procedure. Furthermore, since the set of all regular expressions R not describing Σ^* is, easily seen to be, a nondeterministic csl, we get the following result:

$$\text{DCSL} = \text{NDCSL} \text{ if and only if } L_3 = \{R | R \text{ regular expression}, L(R) \neq \Sigma^*\}$$

is a deterministic context-sensitive language.

Similarly one proves that the family of context-sensitive languages is closed under complementation if and only if $\overline{L}_3$ is a csl.

A generalization of this result leads to a metatheorem about properties of regular expressions which link the LBA problems to the tape complexity of many other decision problems about regular sets.

Let P be any property on the regular sets over $\Sigma = \{0, 1\}$ such that

(1) $P(\Sigma^*) = \text{True}$, and

(2) the set of languages $\bigcup_{x \in \Sigma^*} \{x \backslash L | P(L) = \text{True}\}$ is properly contained in the family of regular sets over Σ where $x \backslash L = \{w | xw \in L\}$.

Let

$$L = \{R | R \text{ is a regular expression over } \{0, 1\} \text{ and } P[L(R)] = \text{False}\}$$

be a nondeterministic csl. Then L is a deterministic csl if and only if DCSL = NDCSL.

Similarly,

$$\widetilde{L} = \{R | R \text{ is a regular expression over } \{0, 1\} \text{ and } P[L(R)] = \text{True}\}$$

is a nondeterministic csl if and only if the family of nondeterministic cls's are closed under complementation.

To illustrate the power of this result we list five other decision problems about regular sets such that any one of them can be recognized by a det lba if and only if NDCSL = DCSL, and furthermore if the complement of any one of these languages is a csl then the context-sensitive languages are closed under complementation. In all examples R and S are restricted regular expressions over $\{0, 1\}$:

$\{(R, S) | L(R) \neq L(S)\}$,
$\{R | L(R) \neq \Sigma^*\}$,
$\{R | L(R) \text{ is coinfinite}\}$,
$\{R | L(R) \neq \text{REVERSAL } L(R)\}$,
$\{R | L(R) \neq L(R^*)\}$.

2. Hardest tape and time recognizable CSL. In this section we give the first of two proofs that there exists a hardest tape and time recognizable context-sensitive language and show, furthermore, that the LBA problem is equivalent to the problem of eliminating nondeterminism in nonwriting automata or flowchart computations.

For the sake of completeness we recall that a *linearly bounded automaton* is a one-tape Turing machine whose input is placed between end markers and the TM cannot go past these end markers. Thus all the computations of the lba are performed on as many tape squares as are needed to write down the input and since the lba can have arbitrary large (but fixed) tape alphabet, we see that the amount of tape for any given lba (measured as length of equivalent binary tape) is linearly bounded by the length of the input word. If the TM defining the lba operates deterministically we refer to the automaton as a *deterministic lba,* otherwise as a *nondeterministic lba* or simply an lba.

Since the connection between linearly bounded automata and context-sensitive languages is well known we will also refer to the languages accepted by nondeterministic and deterministic lba's as nondeterministic and deterministic context-sensitive languages, respectively.

The essence of the first proof is to write down a universal context-sensitive language so that no other csl can be more difficult to recognize. The suprising thing is that this can be done very easily. Below we give a "universal" csl.

$$L_1 = \{\#M_i\# \text{ CODE } (x_1x_2 \cdots x_n)\# \mid x_1x_2 \cdots x_n \text{ is accepted by lba } M_i\}.$$

Thus the sequences in L_1 consist of a simple encoding of an lba, M_i, followed by an encoded form of an input accepted by M_i. The input encoding CODE $(x_1x_2 \cdots x_n)$ is any straightforward, symbol by symbol encoding of sequences over alphabets of arbitrary cardinality (the input and tape alphabet of M_i) into a fixed alphabet, say $\{0, 1, \#\}$, with the provision that $|\text{CODE } (x_j)| \geqslant$ the cardinality of the tape alphabet of M_i.

It is easily seen that L_1 is a csl since it can be accepted by a nondeterministic lba M which simulates M_i on input $x_1 \cdots x_n$. Since M_i uses no more tape than required to write down the input, the encoded input CODE $(x_1 \cdots x_n)$ gives enough tape for M to simulate M_i. Thus L_1 is a context-sensitive language and we get the next result in terms of L_1.

THEOREM 1. 1. *$L_1 \in NDCSL$.*

2. *$L_1 \in DCSL$ iff $NDCSL = DCSL$.*

3. *$\overline{L}_1 \in NDCSL$ iff the family of context-sensitive languages is closed under complementation.*

PROOF. From the construction of L_1 we know that L_1 is a csl. This follows, as mentioned above, since the codes for the input symbols x_j of M_i

are long enough to encode all tape symbols of M_i. Thus NDCSL = DCSL *implies* that L_1 is recognized by a deterministic lba.

On the other hand, if L_1 is recognizable by a deterministic lba M_D then DCSL = NDCSL since for every ndlba M_i we can recursively construct an equivalent deterministic lba $M_{D(i)}$. The dlba $M_{D(i)}$ operates as follows: For input $x_1 \cdots x_n$, $M_{D(i)}$ writes $\#M_i\#\mathrm{CODE}(x_1x_2 \cdots x_n)\#$ on its tape and starts the dlba M_D on this input and accepts the input iff M_D accepts its input. Because of the definition of M_D the input $\#M_i\#\mathrm{CODE}(x_1x_2 \cdots x_n)\#$ is accepted by M_D if and only if the input $x_1 \cdots x_n$ is accepted by M_i and therefore $M_{D(i)}$ and M_i accept the same set. Furthermore, since the length $\#M_i\#\mathrm{CODE}(x_1x_2 \cdots x_n)\#$ is linearly bounded by the length of $x_1x_2 \cdots x_n$ (for any fixed i) we see that $M_{D(i)}$ is a deterministic lba. Thus NDCSL = DCSL, as was to be shown.

The third part of this theorem follows by a similar argument.

It is interesting to note that if L_1 can be recognized on a deterministic lba then all nondeterministic tape computations using $L_i(n) \geq n$-tape can be replaced by equivalent deterministic computations using no more tape. Furthermore, there is a recursive translation which maps the nondeterministic Turing machines onto the equivalent determinisitc Turing machines.

COROLLARY 2. *DCSL = NDCSL if and only if there exists a recursive translation σ such that for every nondeterministic TM M_i, which uses $L_i(n) \geq n$-tape, $M_{\sigma(i)}$ is an equivalent deterministic TM using no more than $L_i(n)$-tape.*

PROOF. The "if" part of the corollary is obvious.

To show the "only if" part, let M_i be any nondeterministic *TM* accepting the set $A_i \subseteq \Sigma^*$ and using $L_i(n) \geq n$-tape. We first define two auxilliary languages used in the proof. Let

$$A_i' = \{\#w\#^t | M_i \text{ on input } w \text{ uses more than } (t + |w| - 1) \text{ tape squares}\}.$$

Clearly, A_i' is a nondeterministic csl, since we can run M_i nondeterministically on input w and see whether for some choice of moves more than $(t + |w| - 1)$-tape is required. But if DCSL = NDCSL then A_i' is accepted by a deterministic lba M_i'.

Next, we define

$$A_i'' = \{\#w\#^t | M_i \text{ accepts } w \text{ using no more than } (|w| + t) \text{ tape squares}\}.$$

Again, A_i'' is accepted by a nondeterministic lba and therefore, by our assumption, A_i'' is accepted by a deterministic lba M_i''.

We now show that from M_i' and M_i'', which can be obtained recursively from

M_i by Theorem 1, we can recursively obtain $M_{\sigma(i)}$ which accepts A_i using no more than $L_i(n)$ deterministic tape.

$M_{\sigma(i)}$ operates as follows:

1. For input $w = x_1 \cdots x_n, M_{\sigma(i)}$ finds the largest t_0 (if it exists) such that $\#w\#^{t_0}$ is in A_i' by successively checking $\#w\#, \#w\#^2, \#w\#^3, \cdots$ with the deterministic lba M_i'.

2. On $\#w\#^{t_0}, M_{\sigma(i)}$ simulates the deterministic lba M_i'' and accepts the input w if M_i'' accepts $\#w\#^{t_0}$.

Clearly, $M_{\sigma(i)}$ accepts A_i on deterministic tape $L_i(n)$, as was to be shown.

From the above results we see that if DCSL = NDCSL then all other deterministic and nondeterministic tape-bounded computations using more than a linear amount of tape are the same. On the other hand, we have not been able to force the equality downward. For example, we have not been able to show that if all deterministic and nondeterministic tape-bounded computations using $L_i(n) \geq 2^n$-tape are the same, then DCSL = NDCSL.

Similarly, it could happen that DCSL = NDCSL but that the $(\log n)$-bounded deterministic languages are properly contained in the nondeterministic $(\log n)$-bounded computations.

Our next result shows that the previous theorem can be generalized to hold for a wide class of tape-bounded languages. Similar results have also been obtained by R. V. Book [1] using AFL theoretic techniques.

We say that $f: N \to N$ is a *semihomogeneous function* if for all $c > 0$ there exists a k_c such that $f(cn) \leq k_c f(n)$. Thus $f(n) = n^5$ is a semihomogeneous function but $f(n) = 2^n$ is not. We say that $f(n)$ is *nondeterministic tape constructible* iff there exsits a nondeterministic TM which for input a^n computes $f(n)$ using no more than $f(n)$ tape squares.

Let

$$L_f = \left\{ \begin{array}{l|l} \#\text{CODE}\,(x_1 x_2 \cdots x_n)\# & |M_i| \leq |\text{CODE}\,(x_1 x_2 \cdots x_n)|, \\ \#M_i\# \cdots \qquad\qquad\qquad \# & x_1 x_2 \cdots x_n \text{ is accepted by } M_i \text{ using} \\ & \text{no more than } f(n)\text{-tape, and} \\ & |\text{CODE}\,(x_j)| \geq \text{cardinality of tape} \\ & \qquad\qquad \text{alphabet of } M_i \end{array} \right\}.$$

We assume that all codes of input and tape alphabet symbols of M_i have the same length.

THEOREM 3. *Let f be a nondeterministic tape constructible, semihomogeneous function such that, for all $k, k \geq 1, f(kn) \geq kf(n) > 0$. Then L_f is nondeterministic $f(n)$-tape recognizable. Furthermore, L_f is deterministic*

$f(n)$-tape recognizable iff the deterministic and nondeterministic $f(n)$-tape recognizable languages are the same.

PROOF. Let L_f be defined as above and note that $f(kn) \geq kf(n)$ implies, for all n, $f(n) \geq c \cdot n$, for a fixed constant $c > 0$.

Then the following algorithm describes an f-tape bounded nondeterministic TM which recognizes L_f.

1. Check $|M_i| \leq |\text{CODE}(x_1 x_2 \cdots x_n)|$.
2. Verify that the format is correct and that the proper coding is used.
3. On a work track of the tape mark off $|\Sigma| f(n)$ squares for scratch space, Σ is the tape alphabet of M_i.
4. Simulate M_i on $x = x_1 x_2 \cdots x_n$ using the scratch space from 3. Accept the input iff M_i accepts x. *Note.* We have enough tape since the simulation needs to encode no more than $f(n)$ tape symbols of M_i.

The space required to execute (1) and (2) is linear in n. To execute steps (3) and (4) we need $|\Sigma| f(n)$ tape squares. But $|\Sigma| f(n) \leq f(|\Sigma| n) \leq k \cdot f(n)$, thus L_f is nondeterministic $f(n)$-tape acceptable.

On the other hand, if L_f is deterministic $f(n)$-tape acceptable, then there exists a deterministic $f(n)$-tape bounded TM M' such that $L(M') = L_f$. We use M' to find for every nondeterministic $f(n)$-tape bounded TM an equivalent deterministic $f(n)$-tape bounded machine. For any TM M_i construct $M_{\sigma(i)}$ as follows:

1. Short inputs are accepted by table look-up. For input $x_1 x_2 \cdots x_n$ such that $|\text{CODE}(x_1 x_2 \cdots x_n)| \geq |M_i|$, $M_{\sigma(i)}$ writes out

$$\begin{array}{l}\#\text{CODE}(x_1 x_2 \cdots x_n)\# \\ \#M_i\# \cdots \qquad\qquad\quad \#\end{array}$$

2. $M_{\sigma(i)}$ applies M' to the new input from (1). The tape required by $M_{\sigma(i)}$ is less than $k_1 n + f(k_1 n)$ which is less than $k_1 n + k_2 f(n)$, since f is semihomogeneous. But then the required tape can be bounded by $cf(n)$ and we see that $M_{\sigma(i)}$ is a deterministic $f(n)$-tape bounded TM, as was to be shown.

Note that in Theorem 3 we could replace the condition $f(kn) \geq kf(n)$ by the weakened condition $f(kn) \geq (\log k) f(n)$, and still carry through the proof. Thus we know, for example, that there exists hardest tape recognizable languages for functions such as: $n^{1/2}$, $n^{1/3}$, $n^{2/3}$, etc. Combining this observation with our previous result we get

COROLLARY 4. *For any positive rational number r the language L_{n^r} is $f(n) = n^r$-nondeterministic tape recognizable. Furthermore L_{n^r} is deterministic*

n^r-tape recognizable iff all n^r nondeterministic tape bounded computations can be so recognized.

So far all considerations have involved tape as our computational complexity measure. It turns out that the hardest tape recognizable language L_1 is also a hardest time recognizable context-sensitive language. We cast our result in terms of polynomial time computable languages.

THEOREM.5. *All context-sensitive languages can be recognized in deterministic polynomial time (nondeterministic polynomial time) if and only if the csl L_1 can be recognized in deterministic polynomial time (nondeterministic polynomial time).*

PROOF. Recall that

$$L_1 = \{\#M_i\#\text{CODE}(x_1x_2 \cdots x_n)\#|x_1x_2 \cdots x_n \text{ is accepted by lba } M_i\}.$$

Clearly, if csl's are accepted in polynomial time then so is the csl L_1.

If L_1 is accepted in polynomial time by a multi-tape Turing machine M then for any lba M_i we can recursively obtain a TM $M_{\rho(i)}$ accepting the same language in polynomial time. $M_{\rho(i)}$ operates as follows: For input in $x_1x_2 \cdots x_n$, $M_{\rho(i)}$ writes down $\#M_i\#\text{CODE}(x_1x_2 \cdots x_n)\#$ and then simulates M · on this input. Clearly, if M operates in polynomial time then so does $M_{\rho(i)}$, as was to be shown.

It is worth mentioning that Greibach [4] has recently exhibited a context-free language which plays the same role among context-free languages as L_1 does for context-sensitive languages. Namely, this context-free language is the hardest time and tape recognizable csl and there also exist two recursive translations mapping context-free grammars onto Turing machines recognizing the language generated by the grammar in the minimal time and on the minimal amount of tape, respectively. Though at this time we do not know what is the minimal time or tape required for the recognition of context-free languages.

Before proceeding with the study of context-sensitive languages we will state two conjectures about tape requirements for the recognition of context-free languages.

CONJECTURE 1. *There exists a context-free language which cannot be recognized nondeterministically on* ($\log n$)*-tape (though we know that all context-free languages are deterministically recognizable on* $[\log n]^2$*-tape* [11]).

CONJECTURE 2. *If* L *is a nonregular context-free language which can be recognized deterministically on* ($\log \log n$)*-tape, then* $\bar{L}$ *is not a context-free language. We know that there exist* ($\log \log n$)*-tape recognizable context-free* *languages* [11], *but in all such cases the complement is not a context-free language and its recognition does not require counting (i.e., (log n)-tape). On the other hand, intuitively it seems that if L and $\overline{L}$ are nonregular context-free languages then the recognition process must involve counting and therefore must require at least* (log n)-*tape.*

Finally we note that the methods used to construct the "universal" csl L_1 can be used to construct other "universal" languages. We illustrate this by constructing the language $\widetilde{L}_1$, which plays the same role for nondeterministic polynomial time-bounded computations as L_1 does for the context-sensitive languages.

Let DPTIME and NDPTIME designate the families of languages accepted by deterministic and nondeterministic polynomial time-bounded Turing machines, respectively.

We will say that a language L is *p-complete* iff L is in NDPTIME and for all L_i in NDPTIME there exists a deterministic polynomial time-bounded function f_i such that

$$x \text{ is in } L_i \text{ iff } f_i(x) \text{ is in } L.$$

Let $\widetilde{L}_1 = \{\#M_i\#\text{CODE}\,(x_1x_2 \cdots x_n)\#^{3|M_i|t} \mid x_1x_2 \cdots x_n$ is accepted by the one-tape, nondeterministic TM M_i in time $t\}$.

THEOREM 6. *The language $\widetilde{L}_1$ is accepted in nondeterministic linear time by a four-tape TM. Furthermore, $\widetilde{L}_1$ is in DPTIME iff NDPTIME = DPTIME.*

PROOF. It is easily seen that a four-tape TM M' can accept $\widetilde{L}_1$ in linear time. We indicate how M' uses its tapes: On the first sweep of the input M' checks the format of the input, copies M_i from the input on the first working tape and $\#^{3|M_i|t}$ on the second working tape. The third working tape is used to record the present state of M_i (in a tally notation) during the step-by-step simulation of M_i. It is seen that with the available information on its working tapes M' can simulate M_i on the input in time $2|M_i|t$ (for an appropriate, agreed upon representation of M_i). Thus M' operates in nondeterministic linear time and accepts $\widetilde{L}_1$. Therefore, $\widetilde{L}_1$ is in NDPTIME and the assumption that NDPTIME = DPTIME implies that $\widetilde{L}_1$ is in DPTIME.

To prove that $\widetilde{L}_1$ in DPTIME implies that DPTIME = NDPTIME, assume that $\widetilde{L}_1$ is accepted by a deterministic TM M'' operating in deterministic time n^p. Then for any nondeterministic TM M_i working in time n^q we can recursively construct a TM $M_{\sigma(i)}$ operating in deterministic polynomial time as follows:

1. For input $x_1x_2 \cdots x_n$, $M_{\sigma(i)}$ writes down

$$\#M_i\#\text{CODE}\,(x_1x_2 \cdots x_n)\#^{3|M_i|n^q}.$$

2. $M_{\sigma(i)}$ starts the deterministic machine M'' on the sequence in (1) and accepts the input $x_1x_2 \cdots x_n$ iff M'' accepts its input.

Clearly, M_i and $M_{\sigma(i)}$ are equivalent; furthermore $M_{\sigma(i)}$ operates deterministically in time less than

$$2[3|M_i|n^q + |\#M_i\#\text{CODE}\,(x_1x_2 \cdots x_n)|]^p \leqslant Cn^{pq}.$$

Thus $M_{\sigma(i)}$ operates in deterministic polynomial time, as was to be shown.

The previous proof shows that if $\widetilde{L}_1$ is in DPTIME, then we can recursively obtain for every M_i running in time n^q an equivalent deterministic TM running in time $O[n^{pq}]$. Unfortunately, for a given TM we cannot recursively determine the running time and thus we do not know whether M_i runs in polynomial time or not. Even if we know that M_i runs in polynomial time we can still not recursively determine the degree of the polynomial.

Our next result shows that, nevertheless, we can get a general translation result. For a related result see [3].

THEOREM 7. *DPTIME = NDPTIME iff there exists a recursive translation σ and a positive integer k, such that for every nondeterministic TM M_i, which uses time $T_i(n) \geqslant n$, $M_{\sigma(i)}$ is an equivalent deterministic TM working in time $O[T_i(n)^k]$.*

PROOF. The "if" part of the proof is obvious. To prove the "only if" part assume that DPTIME = NDPTIME. We will outline a proof that we can recursively construct for any M_i, running time $T_i(n) \geqslant n$, an equivalent deterministic TM $M_{\sigma(i)}$ operating in time $O[T_i(n)^k]$, for a fixed k.

In our construction we use two auxillary languages:

$$B_i' = \{\#w\#^t | M_i \text{ accepts } w \text{ in less than } t \text{ time}\},$$

$$B_i'' = \{\#w\#^t | M_i \text{ on input } w \text{ takes more than } t \text{ time}\}.$$

Clearly, both languages can be accepted in nondeterministic linear time. Therefore, by our previous result, we can recursively construct two deterministic machines M_i' and M_i'' which accept B_i' and B_i'', respectively, and operate in time $O[n^p]$. From M_i' and M_i'' we can recursively construct the deterministic TM $M_{\sigma(i)}$, which operates as follows:

1. For input w, $M_{\sigma(i)}$ finds the smallest t_0 such that $\#w\#^{t_0}$ is not in B_i''. This is done by checking with M_i'' successively $\#w\#, \#w\#^2, \#w\#^3, \cdots$.

2. $M_{\sigma(i)}$ starts M_i' on input $\#w\#^{t_0}$ and accepts w iff M_i' accepts $\#w\#^{t_0}$.

Clearly, $M_{\sigma(i)}$ is equivalent to M_i and $M_{\sigma(i)}$ operates in time

$$O\left[\sum_{1}^{T_i(n)} n^p\right] = O[T_i(n)^{p+1}].$$

By setting $k = p + 1$, we have completed the proof.

We conclude by observing that $\widetilde{L}_1$ is a p-complete problem, as defined above.

3. Nonwriting devices and flowcharts. Next we will show that the LBA problem is equivalent to problems about eliminating nondeterminism in some very simple nonwriting automata. Then we will use this result to show that the LBA problem is also equivalent to eliminating nondeterminism from a single 10-variable elementary flowchart by using more variables.

A *k-head finite automaton* (or a *multi-head finite automaton*) is a one-tape Turing machine with k read-only heads, $k = 1, 2, 3, \cdots$. The input string is written on the tape with special end markers at both ends of the input, and the finite automaton is so designed that the read heads cannot leave the input. The automaton is an accepting device and an input is accepted if, after starting the automaton in its starting state with all heads on the left end marker, the automaton enters an accepting state and halts. We assume that the automaton is capable of sensing when two heads are on the same tape square. We distinguish between deterministic and nondeterministic multi-head automata.

We first establish a relationship between linearly bounded languages and (log n)-tape bounded languages over one-letter alphabets, due to Savitch **[15]**.

For any language A over an alphabet $\Sigma = \{a_1, a_2, \cdots, a_k\}$, $A \subseteq \Sigma^*$, let

$$\text{TALLY}(A) = \{1^{n(w)} \mid w \text{ in } A\},$$

where n maps each word w in Σ^* onto the number $n(w)$ which w denotes in k-adic notation, that is,

$$n(a_{i_0} a_{i_1} \cdots a_{i_t}) = \sum_{j=1}^{t} a_{i_j} k^j$$

(where we interpret a_i as i).

Clearly, this mapping establishes a one-one correspondence between strings over Σ and nonnegative integers; zero is denoted by the null string.

LEMMA 8. *The language* A, $A \subseteq \Sigma^*$ *with* $|\Sigma| = k$, *is accepted by a deterministic (nondeterministic) linearly bounded automaton if and only if* TALLY (A) *is accepted by a deterministic* (*nondeterministic*)(log n)*-tape bounded Turing machine.*

PROOF. Since going from A to TALLY (A) the length of every string is increased exponentially, for input 1^{n_i} the $(\log n_i)$-tape bounded Turing machine has as much tape available as the lba has for input n_i. Thus a $(\log n)$-tape bounded Turing machine can accept TALLY (A) if an lba can accept A. Conversely, if $A = \{1^{n_i}\}$ is accepted by a $(\log n)$-tape bounded TM, then $\{n_i\}$, where n_i is written in k-adic notation with $k \geqslant 2$, can be accepted by an lba which simulates the $(\log n)$-tape bounded TM. Since this lba has enough tape to carry out the simulation we have the desired result.

Thus we immediately obtain the following result.

COROLLARY 9. *The deterministic and nondeterministic context-sensitive languages are the same if and only if the deterministic and nondeterministic* $(\log n)$*-bounded languages over one-letter alphabets are the same.*

At the same time it is known that

LEMMA 10. *The language* $A, A \subseteq \Sigma^*$, *is accepted by a deterministic (nondeterministic) multi-head finite automation if and only if* A *is accepted by a deterministic (nondeterministic)* $(\log n)$*-tape bounded Turing machine.*

PROOF. (For a more complete proof see [5].) The basic idea of the proof is that a $(\log n)$-tape bounded TM can count up to n (k-times) and thus can encode the k-head positions of a k-head automaton, say in binary form, on the $(\log n)$-tape and use this encoding for a stepwise simulation of the k-head finite automaton. Thus every set accepted by a k-head automaton is also accepted by a $(\log n)$-tape bounded TM.

Conversely, every $(\log n)$-tape bounded Turing machine can be simulated by a k-head finite automaton which encodes the tape content of the $(\log n)$-tape bounded Turing machine by its head positions on the input tape. Since on a $(\log n)$-tape we can record no more than n^p different patterns (for some p), we see that on input of length n, p heads can encode all these patterns. With a few additional bookkeeping heads, utilizing the encoded $(\log n)$-tape bounded TM tape content, the k-head automaton can simulate the $(\log n)$-tape bounded TM. Thus every $(\log n)$-tape bounded language can be accepted by a multi-head automaton. Since these considerations hold for deterministic as well as nondeterministic automata, we have completed the outline of the proof.

From this we get Savitch's result.

COROLLARY 11. *The deterministic and nondeterministic context-sensitive languages are the same if and only if the languages over a one-letter alphabet accepted by the deterministic and nondeterministic multi-head finite automata are the same.*

Next we show that we can strengthen this result by using the language TALLY (L_1), where L_1 is the "universal" csl defined before.

THEOREM 12. 1. *The language* TALLY (L_1) *is recognizable by a* k_0*-head nondeterministic automaton.*

2. TALLY (L_1) *is recognizable by a deterministic* $(k_0 + p)$*-head automaton iff DCSL = NDCSL.*

PROOF. Since L_1 is a ndcsl we know, from our previous results, that TALLY (L_1) is accepted by a k_0-head nondeterministic automaton. (k_0 can be explicitly computed from L_1.)

Similarly, if TALLY (L_1) can be accepted by a deterministic $(k_0 + p)$-head automaton then we know that L_1 can be accepted by a dlba, and vice versa. But then, using Theorem 1, we get that TALLY (L_1) is deterministically recognizable on some $(k_0 + p)$-head automaton iff DCSL = NDCSL, as was to be shown.

The next result shows that the number of heads k_0 in the previous result can be reduced to 7 heads. For $L \subseteq a^*$ define $L^{[k]} = \{a^{n^k} | a^n \text{ in } L\}$.

COROLLARY 13. 1. *The language* $[\text{TALLY}(L_1)]^{[k_0]}$ *is accepted by a* 7-*head nondeterministic finite automaton.*

2. $[\text{TALLY}(L_1)]^{[k_0]}$ *is accepted by a deterministic mulit-head automaton iff DCSL = NDCSL.*

The proof follows from the next lemma.

LEMMA 14. *Let* $A, A \subseteq a^*$, *be a set accepted by a nondeterministic k-head finite automaton. Then* $A^{[k]} = \{a^{n^k} | a^n \text{ in } A\}$ *is accepted by a 7-head nondeterministic finite automation and* A *is accepted by a deterministic multi-head finite automaton if and only if* $A^{[k]}$ *is accepted by a deterministic multi-head finite automaton.*

PROOF. The main tool in this proof is the method of encoding the position of the k-heads of a finite automaton M on the input a^n by one head of an automaton M_1 an input a^{n^k} and then using six additional read-only heads to carry out a simulation of M by M_1. The essential steps in the simulation are described below. First we note that with five read heads a deterministic finite automaton can check whether the input a^t is such that $t = n^k$, for some n. Thus the format of the input can be checked and a head can be placed on the nth tape square if $t = n^k$.

To encode the k heads of M on input a^n as one head position of M_1 on input a^{n^k}, order the k heads arbitrarily and place the "encoding" head of

M_1 on the rth tape square $1 \leqslant r \leqslant n^k$ with

$$r - 1 = (d_1 - 1) + (d_2 - 1)n + (d_3 - 1)n^2 + \cdots + (d_k - 1)n^{k-1}$$

iff the ith head of M, $1 \leqslant i \leqslant n$, is on the d_ith tape square. After this by a lengthy but straightforward argument one can show that M_1 can carry out a step by step simulation of M and thus M accepts input a^n if and only if M_1 accepts input a^{n^k}.

Clearly, if A can be accepted by a deterministic multi-head finite automaton then so can $A^{[k]}$ for any k. If $A^{[k]}$ can be accepted by a deterministic p-head finite automaton M_2 then we can design a $p(k + 1)$-head deterministic automaton M_3 which accepts A.

For input a^n the automaton M_3 will simulate M_2 on input a^{n^k} as follows: M_3 uses the first p heads to mimic the p heads of M_2, as long as these heads stay on the first n tape squares. If a head of M_2 goes further than the first n tape squares (recall that the simualted input is n^k long) then k heads are used on the input of length n to count how far the head has moved. Since we can count up to n^k with k heads on an input of length n, the $(k + 1)p$ heads suffice for M_3 on input a^n to simulate M_2 on input a^{n^k}. Thus M_3 accepts a^n if and only if M_2 accepts a^{n^k}, but then M_3 accepts A. Thus $A^{[k]}$ is a deterministic language if and only if A is. This completes the proof.

Next we show that the previous results have a natural interpretation for flowchart computations, thus relating the classic nondeterminism problem for context-sensitive languages to a somewhat more programming oriented problem.

We say that a flowchart is *elementary* (or an *E-flowchart*) if and only if it is a flowchart made up of the assignment statements

$$x := x - 1,$$

$$x := y,$$

and the tests

$$x = 0,$$

$$x = y.$$

An E-flowchart is *deterministic* if and only if every assignment statement and every test branch leads to exactly one assignment statement or test. If some assignment statement or test branch leads to more than one assignment or test, or leads to one or more assignments and tests, then the flowchart is *nondeterministic*.

The flowcharts are used as accepting devices of sets of integers. The integer n is accepted if and only if the flowchart computation with the first variable set equal to n ends at some exit labelled with "accept." Otherwise the input is rejected. (Note that the accepting condition can be handled in many different ways. For example, we could have demanded that the computation halts and that a specified variable is set to one for accepting and zero for rejecting.)

THEOREM 15. *There exists a set of integers L accepted by a nondeterministic E-flowchart with* 10 *variables such that the following two statements are equivalent.*

1. *L is accepted by a deterministic E-flowchart.*
2. *DCSL = NDCSL.*

PROOF. The proof consists of a reasonably straightforward simulation of multi-head automata by E-flowcharts and vice versa.

In simulating the flowcharts on k-head automata the head positions on the input tape encode the contents of the variables of the flowchart and vice versa. The three additional variables are needed to obtain a subflowchart which performs the assignment $x := x + 1$ for x less than the input variable and to permanently store the input. This completes the outline of the proof.

For related results see Warkentin and Fischer [16].

We do not know whether the number of heads or the number of flowchart variables can be reduced further in the two previous results. We conjecture, however, that this is the case. We believe that it would be worthwhile to investigate the nondeterministic k-head automata languages over a one-symbol alphabet for $k = 2$ and 3. The case of 2 heads seems simple and it would be very interesting to determine whether all 2-head nondeterministic finite automata can be replaced by equivalent deterministic multi-head finite automata. It is our hope that these k-head automata with small values of k may provide a place where some further insights can be gained into the LBA problem and, more generally, into the nature of nondeterminism in computing.

We also believe that the linearly bounded automata with oracles deserve further investigation. The main problem here is to determine whether there exist recursive oracles such that the deterministic and nondeterministic lba language accepted with these oracles are different.

It is interesting to note that T. Baker [2] has shown that there are recursive oracles for which the deterministic and nondeterministic polynomial time-bounded TM computations are the same and that there are other oracles for which they are different.

4. Decision problems about regular expressions. In this section we show that the LBA problem can be related to several natural decision problems about regular expressions. This approach yields many hardest tape recognizable languages which appear more natural than the "universal" context-sensitive language constructed in the second section.

The main tool in this work is the observation made by Meyer and Stockmeyer [12] that restricted regular expressions can be used to describe invalid lba computations very economically. Thus these results, as well as many other results about the complexity of various decision problems ([6], [7], [12]), should also be viewed as results about the descriptive power of regular expressions.

A *restricted regular expression* or simply a *regular expression* is any valid expression over the alphabet consisting of $\Sigma, \cdot, +, ^*$, and the delimitors (,). The operators $\cdot, +$ and * have their well-known meaning of *concatenation, set union* and *Kleene closure.* For a regular expression R the set of sequences described by R is designated by $L(R)$.

Next we look at *valid lba computations* which will be used to link the LBA problem to the complexity of several decision problems about regular expressions. Consider an lba M with tape alphabet T and state set Q working on an input $y = x_1 \cdots x_n$. At each discrete time interval during the computation we can describe the state of the computation by giving the tape content, the head position of M and its state. If the computation is deterministic then after k steps of computing there will be a unique configuration describing the situation and for a nondeterministic lba there will be a set of possible configurations. To make these ideas more precise we will refer to a sequence

$$x_1x_2 \cdots x_{j-1}(q, x_j)x_{j+1} \cdots x_n$$

as an *instantaneous description.* This sequence means that the tape content is $x_1 \cdots x_n$, the lba is in state q and the reading head is scanning the jth tape symbol x_j. Thus an instantaneous description is any string in $[T + (Q \times T)]^*$, which contains exactly one symbol in $Q \times T$. If the start state is q_0 then $(q_0, x_1)x_2 \cdots x_n$ is an *initial configuration* and any configuration containing a halting state is a *final configuration.* One instantaneous description ID_{i+1} *follows* ID_i if and only if there exists a move of M which changes ID_i in one operation to ID_{i+1}. A *valid computation* of M on input $y = x_1 \cdots x_n$ is a sequence of instantaneous descriptions

$$\#\mathrm{ID}_0 \# \mathrm{ID}_1 \# \mathrm{ID}_2 \# \cdots \mathrm{ID}_t \#$$

where ID_0 is the initial configuration on the input $x_1 \cdots x_n$, i.e.,

$$\mathrm{ID}_0 = (q_0, x_0)x_2x_3 \cdots x_n,$$

ID_t is a final configuration and for all $i, 0 \leqslant i < t, \text{ID}_{i+1}$ follows ID_i. We denote the set of all valid computations of M on input y by $R_M(y)$. Thus we see that y is accepted by M if and only if $R_M(y) \neq \varnothing$ or, equivalently, the set of *invalid computations* of M and $y, \overline{R}_M(y)$, must not contain all sequences, i.e.,

$$I_M(y) = \overline{R}_M(y) \neq \Sigma^*.$$

The main observation [due to Meyer and Stockmeyer] is that, for every lba M and input y, the set of invalid computations of M on y is a regular set and that it can be described by a restricted regular expression such that

$$|I_M(y)| \leqslant c_M |y|,$$

and furthermore that a deterministic lba can map y onto $I_M(y)$. This is the critical step in the argument which links lba computations to regular expressions.

Thus we have the following result:

THEOREM 16 (MEYER AND STOCKMEYER). *Let M be a nondeterministic lba with tape symbol set T, state set Q, and set of designated accepting states F, with $F \subseteq Q$. Let all accepting states be final. Let q_0 be the unique start state of M. Let $y = x_1 x_2 \cdots x_n$ be an input to M.*

Then there is a deterministic lba M' such that M', starting with $\# x_1 x_2 \cdots x_n \#$ on its tape, halts with a regular expression β_y over Σ on its tape such that $L(\beta_y) = \overline{R_M(y)} = I_M(y)$.

PROOF. We only sketch the idea of the proof. For a complete proof see [6].

β_y is the union of:

β_1 the set of strings that do *not* begin with $\#(q_0, x_1)x_2 \cdots x_n \#$;

β_2 the set of strings that do not contain a symbol (q_f, t), where $q_f \in F$;

β_3 the set of words that make a mistake between one i.d. and the next (i.e., ID_{j+1} does not follow from ID_j by one application of a move rule of of M).

But

$$\beta_1 = [(\Sigma - \#) \cup \# \cdot [(\Sigma - (q_0, x_1)) \cup (q_0, x_1) \cdot [(\Sigma - x_2) \cup x_2 \cdot [\cdots \cup x_n \cdot [(\Sigma - \#)] \cdots]]]] \cdot \Sigma^*.$$

The reader should note the similarity of the above to Horner's method for evaluating polynomials, i.e.,

$$a_0 + a_1x + a_2x^2 + \cdots + a_nx^n = a_0 + x[a_1 + x[a_2 + \cdots + x[a_n] \cdots]],$$

$$\beta_2 = \left[\Sigma - \left(\bigcup_{q_f \in F} \{q_f\} \times T\right)\right]^*,$$

$$\beta_3 = \bigcup_{\sigma_1, \sigma_2, \sigma_3 \in \Sigma} \Sigma^* \cdot \sigma_1 \cdot \sigma_2 \cdot \sigma_3 \cdot \Sigma^{|y|-2} \cdot [\Sigma^3 - f_M(\sigma_1, \sigma_2, \sigma_3)] \cdot \Sigma^*,$$

where $f_M: \Sigma^3 \to 2^{\Sigma^3}$, which maps correct triples of symbols into correct triples. Essentially β_3 says that mistakes occur n symbols apart.

The remainder of the proof consists in noting that $|\beta_y| \leq C_M \cdot |y|$ and that given y, the time required to deterministicaly write out β_y is bounded by a polynomial in $|\beta_y|$.

Using Theorem 16 a simple coding argument yields the following:

THEOREM 17. *Let*

$$L_4 = \{(R_1, R_2) | R_1 \text{ and } R_2 \text{ are regular expressions over } \{0, 1\} \text{ and } L(R_1) \neq L(R_2)\}.$$

Then $L_4 \in NDCSL$ *and* $L_4 \in DCSL$ *iff* $NDCSL = DCSL$.

PROOF. If L_4 is in DCSL then we can check $L(\beta_y) \neq \Sigma^*$ on a deterministic lba and from Theorem 16 it follows that $\text{DCSL} = \text{NDCSL}$.

To see that L_4 is in NDCSL, we note that to verify $L(R_1) \neq L(R_2)$ we need only to give a string x one symbol at a time and verify that

$$x \in [L(R_1) \cap L(\bar{R}_2)] \cup [L(\bar{R}_1) \cap L(R_2)].$$

This can be done on a nondeterministic lba in a straightforward way, which completes the proof.

We next extend Theorem 17 to prove a metatheorem about the deterministic tape complexity of many decision problems about the regular sets. Define

$$x \backslash L = \{w | xw \in L\} \quad \text{and} \quad L/x = \{w | wx \in L\}.$$

THEOREM 18. *Let* P *be any predicate on the regular sets over* $\{0, 1\}$ *such that*

1. $P(\{0, 1\}^*)$ *is True, and*
2. $P_L = \bigcup_{x \in \{0,1\}^*} \{x \backslash L | P(L) = True\}$ [*or* $P_R = \bigcup_{x \in \{0,1\}^*} \{L/x | P(L) = True\}$] *is not the set of all regular sets over* $\{0, 1\}$.

Then $\{R \mid R$ *is a regular expression over* $\{0, 1\}$ *and* $P[L(R)] = False\}$ *in DCSL implies NDCSL = DCSL.*

Similarly, $\{R \mid R$ *is a regular expression over* $\{0, 1\}$ *and* $P[L(R)] = True\}$ *in NDCSL implies that NDCSL is closed under complementation.*

PROOF. Let L_0 be a regular set over $\{0, 1\}$ not in $\mathcal{P}_L$. Let $h_0(0) = 00$ and $h_0(1) = 01$. Then given R_i a regular expression over $\{0, 1\}$ we can effectively find in linear space and deterministic polynomial time in $|R_i|$ a regular expression R_j such that

$$L(R_j) = h_0(L(R_i)) \cdot 10 \cdot (0 + 1)^* + (00 + 01)^* \cdot 10 \cdot L_0 + \overline{(00 + 01)^* \cdot 10 \cdot (0 + 1)^*}$$

$$= h_0(L(R_i)) \cdot 10 \cdot (0 + 1)^* + (00 + 01)^* \cdot 10 \cdot L_0 + (00 + 01)^* [\Lambda + 0 + 1 + 11(0 + 1)^*].$$

Case 1. $L(R_i) = (0 + 1)^*$. Then

$$h_0(L(R_i)) = (00 + 01)^* \quad \text{and} \quad L(R_j) = (0 + 1)^*.$$

Hence, $P(L(R_j)) = \text{True}$.

Case 2. $L(R_i) \neq (0 + 1)^*$. Then $\exists x \in (0 + 1)^* - L(R_i)$. Hence $h_0(x) \in (00 + 01)^* - h_0(L(R_i))$. But

$$P(L(R_j)) = \text{True implies } h_0(x)10 \backslash L(R_j) = L_0 \in \mathcal{P}_L.$$

Hence $P(L(R_j))$ is False. Therefore, $P(L(R_j)) = \text{True}$ if and only if $L(R_i) = (0 + 1)^*$.

Thus if $P(L(R_i)) = \text{False}$ is decidable by a dlba then so is $L(R_i) \neq (0 + 1)^*$, and, therefore, by our previous results it follows that NDCSL = DCSL, as was to be shown.

COROLLARY 19. *DCSL = NDCSL iff any one of the following languages is in DCSL. Similarly, NDCSL is closed under complementation iff the complement of any one of the following languages is in NDCSL:*

1. $\{R \mid R$ *is a regular expression and* $L(R) \neq \{0, 1\}^*\}$;
2. $\{R \mid R$ *is a regular expression and* $L(R) \neq L(R^*)\}$;
3. $\{R \mid R$ *is a regular expression and* $L(R) \neq L(R)^{\mathrm{REV}}\}$;
4. $\{R \mid R$ *is a regular expression and* $L(R)$ *is coinfinite*$\}$;
5. $(\forall k \geq 1)$ $\{R \mid R$ *is a regular expression and* $L(R)$ *is not* k*-definite*$\}$.

It is interesting to note that in the proofs of Theorem 16 and 17 we only used regular expressions of star-height 1 (i.e., no nested *'s). Thus if there exists a regular expression R_0 of star-height 1 not in $\mathcal{P}_L$, then Theorem 18 can be changed to read "$\{R|R$ is a regular expression over $\{0, 1\}$ of star-height 1 and $P[L(R)] = \text{False}\}$ in DCSL implies DCSL = NDCSL." We also note that all the languages in Corollary 19 can be chosen to be of star-height 1. Thus we get, for example, the following:

COROLLARY 20. *The language* $\{R|R$ *is a regular expression of star-height* 1 *and* $L(R) \neq L(R^*)\}$ *is a tape and time hardest csl.*

It is natural to ask if there are other regular sets R_0 besides $\{0, 1\}^*$ for which $\{R|R$ is a regular expression and $L(R) \neq R_0\}$ is a tape and time hardest csl. We say that a language L is *bounded* iff there exist strings $w_1, \cdots, w_k$ such that $L \subseteq w_1^* \cdots w_k^*$. If L is not bounded, then L is said to be unbounded. For a regular set R_0 over $\{0, 1\}$, R_0 is unbounded iff there exist strings $r, s, x, y \in \{0, 1\}^*$ such that $R_0 \supseteq r \cdot (0x + 1y)^* \cdot s$ (see [17]).

THEOREM 21. *For all unbounded regular sets* R_0 *over* $\{0, 1\}$, $\{R|R$ *is a regular expression and* $L(R) \neq R_0\}$ *is a tape and time hardest csl.*

PROOF. We only sketch the proof. Since R_0 is an unbounded regular set over $\{0, 1\}$, there exist strings $r, s, x, y \in \{0, 1\}^*$ such that $R_0 \supseteq r(0x + 1y)^*s$. For all regular expressions R_i, a regular expression R_i' can be constructed such that

$$L(R_i') = r \cdot h(L(R_i)) \cdot s + R_0 \cap \sim r \cdot (0x + 1y)^* \cdot s$$

where h is the homomorphism defined by $h(0) = 0x$ and $h(1) = 1y$. But $L(R_i') = R_0$ iff $L(R_i) = \{0, 1\}^*$. Since this construction requires only linear space and deterministic polynomial time in $|R_i|$ on a deterministic lba, the theorem follows.

From the above observations we know that even if we restrict ourselves to regular expressions of star-height 1, the language $\{(R_i, R_j)| L(R_i) \neq L(R_j)\}$ is a hardest tape and time csl. If Kleene's star is dropped completely, we get a *p*-complete problem.

PROPOSITION 22. *Let* R_i, R_j *be regular expressions over* $0, 1, +, \cdot$. *Then* $L = \{(R_i, R_j)| L(R_i) \neq L(R_j)\}$ *is a p-complete language.*

PROOF. See [6].

For other related results see [18].

5. NDPTIME parallels. In this section several parallels of the results in §§2 and 3 are presented. Let PSPACE denote the family of languages accepted by deterministic or nondeterministic polynomially tape bounded TM's. Our first result is an analogue of Corollary 2.

THEOREM 23. *DPTIME (NDPTIME) = PSPACE iff there exists a recursive translation* σ *and a positive integer* k, *such that, for every nondeterministic TM* M_i, *which uses tape* $L_i(n) \geqslant n$, $M_{\sigma(i)}$ *is an equivalent deterministic (nondeterministic) TM working in time* $O[L_i(n)]^k$.

PROOF. If M_i' and M_i'' in the proof of Corollary 2 are deterministic (nondeterministic) polynomially time bounded TM's, then so is $M_{\sigma(i)}$.

Similarly the following parallel of Corollary 4 holds.

PROPOSITION 24. *For all positive rationals* r, *the language* $L_{n^r} \in$ *DPTIME (NDPTIME) iff DPTIME (NDPTIME) = PSPACE.*

The proof of Proposition 24 is obvious and is left for the reader.

Next we present a parallel of 8–12. Let $\mathrm{DPTIME_{sla}}$ ($\mathrm{NDPTIME_{sla}}$) denote the family of languages over $\{1\}$ accepted by deterministic (nondeterministic) polynomial time bounded TM's. Let DEXP (NDEXP) denote the union over all positive integers k of the families of languages accepted by 2^{kn} deterministic (nondeterministic) time bounded TM's.

THEOREM 25. (1) $\mathit{DPTIME}_{sla} = \mathit{NDPTIME}_{sla}$ *iff DEXP = NDEXP.*
(2) $\mathit{NDPTIME}_{sla}$ *is closed under complementation iff NDEXP is.*

PROOF. The proof is based on the properties of the TALLY function introduced in §3. Let $w = w_0 \cdots w_n \in \{1, 2\}^+$. Let TALLY $(w) = \Sigma_{j=0}^n w_j \cdot 2^j$; then TALLY is a bijection from $\{1, 2\}^+$ to $\{1\}^+$. For all $w \in \{1, 2\}^+$, $2^{|w|} - 1 \leqslant |\mathrm{TALLY}(w)| \leqslant 2^{|w|+1} - 2$. For all $y \in \{1\}^+$,

$$|\mathrm{TALLY}^{-1}(y)| - 1 \leqslant \log(|y|) \leqslant |\mathrm{TALLY}^{-1}(y)| + 1.$$

The remainder of the proof is similar to that of Lemma 8 and is left to the reader.

Finally we note that there are hardest time recognizable languages in $\mathrm{NDPTIME_{sla}}$. It can be shown that there exists a language $L_0 \in$ NDEXP such that for all $L_i \in$ NDEXP there exists a deterministic polynomial time bounded function f_i for which $x \in L_i$ iff $f_i(x) \in L_0$ and $|f_i(x)| \leqslant c_i|x|$, where c_i depends only upon f_i. (A construction analogous to that of L_1 in the proof of Theorem 1 yields such a language. Another such language is mentioned in [20].) For any such L_0 let $\mathsf{L}_0 = \mathrm{TALLY}(L_0)$. Then $\mathsf{L}_0 \in \mathrm{NDPTIME_{sla}}$ and for all $\mathsf{L}_i \in \mathrm{NDPTIME_{sla}}$ there exists a deterministic polynomially time bounded function f_i for which $x \in \mathsf{L}_i$ iff $f_i(x) \in \mathsf{L}_0$. Thus L_0 is a hardest time recognizable language in $\mathrm{NDPTIME_{sla}}$. It is difficult to see how L_0 can be p-complete. We feel that L_0 is a good candidate for a language in NDPTIME that is both not in DPTIME and not p-complete.

6. DFA and log n tape-bounded TM's. An analogue of the results in §4 is presented for deterministic finite automata (DFA). Let D log n (ND log n) denote the families of languages accepted by deterministic (nondeterministic) log n tape-bounded TM's. Our results are based on the theorem due to Jones [19]: $\{M \mid M$ is a DFA and $L(M) \neq \varnothing\}$ is a hardest tape language in ND log n.

Following Jones a language L_0 is said to be *log-reducible* to a language L_1, written $L_0 \leqslant_{\log} L_1$, if there is a function f computable by a deterministic off-line TM M_f such that $x \in L_0$ iff $f(x) \in L_1$, where

(a) M_f has a two-way read-only input tape, a two-way read/write work tape, and a one-way write only output tape;

(b) M_f, given any input word x, halts with $f(x)$ on its output tape, and

(c) M_f never scans more than $O(\log(|x|))$ squares on its work tape.

We note that if L_1 is acceptable by some deterministic log n tape-bounded TM, then so is L_0.

PROPOSITION 26 (JONES [19]). *ND* log $n \leqslant_{\log} \{M \mid M$ *is a DFA and* $L(M) \neq \varnothing\}$.

PROOF. See [19].

As an immediate corollary we note that ND log $n \leqslant_{\log} \{M \mid M$ is a DFA and $L(M) \neq \{0, 1\}^*\}$ as well.

THEOREM 27. *Let P be any nontrivial predicate on the regular sets over $\{0, 1\}$ such that $P_{\text{left}} = \{L' \mid L' = x \backslash L, x \in \{0, 1\}^+, P(L)$ is true$\}$ or $P_{\text{right}} = \{L' \mid L' = L/x, x \in \{0, 1\}^+, P(L)$ is true$\}$ is not equal to the set of all regular sets over $\{0, 1\}$. Then ND* log $n \leqslant_{\log} \{M \mid M$ *is a DFA and* $P(L(M))$ *is false*$\}$. *Similarly,* $\{M \mid M$ *is a DFA and* $P(L(M))$ *is true*$\} \in$ *ND* log n *implies that ND* log n *is closed under complementation.*

PROOF. We only sketch the proof. Let L_f be a regular set over $\{0, 1\}$ not in P_{left}.

Case I. P is true for some R_0, where $R_0 \supseteq r \cdot (0 + 1)^*$ for some $r \in \{0, 1\}^*$. For all DFA M_i a DFA M_i' can be constructed such that

$$L(M_i') = r \cdot h(L(M_i)) \cdot 10 \cdot (0 + 1)^* + r \cdot (00 + 01)^* \cdot 10 \cdot L_f$$
$$+ R_0 \cap \sim r \cdot (00 + 01)^* \cdot 10 \cdot (0 + 1)^*,$$

where h is the homomorphism defined by $h(0) = 00$ and $h(1) = 01$. Then $P(L(M_i'))$ is true iff $L(M_i) = \{0, 1\}^*$.

Case II. For no set R_0 and string r are both $P(R_0)$ true and $R_0 \supseteq r \cdot (0 + 1)^*$. Let $P(R_1)$ be true. For all DFA M_i, a DFA M_i' can be constructed such that $L(M_i') = L(M_i) \cdot (0 + 1)^* + R_1$. Then $P(L(M_i'))$ is true iff $L(M_i) = \varnothing$.

In both cases M_i' can be constructed from M_i by a deterministic log n tape-bounded transducer satisfying conditions (a), (b), and (c) above.

We note one corollary analogous to Corollary 19.

COROLLARY 28. *$D \log n = ND \log n$ iff any one of the following languages is in $D \log n$. Similarly, $ND \log n$ is closed under complementation iff the complement of any one of the following languages is in $ND \log n$:*

1. *For all regular sets R_0 over $\{0, 1\}$, $\{M | M$ is a DFA and $L(M) \neq L(R_0)\}$;*
2. *$\{M | M$ is a DFA and $L(M)$ is cofinite$\}$; and*
3. *$(\forall k \geq 1)$ $\{M | M$ is a DFA and $L(M)$ is not k-definite$\}$.*

Finally the conclusion of Theorem 18 can be strengthened as follows: $\{R | R$ is a regular expression over $\{0, 1\}$ and $P(L(R)) = \text{False}\} \geq_{\log}$ NDCSL. (See [18].) Using known results about nondeterministic tape hierarchies, the following holds:

PROPOSITION 29. *For all unbounded regular sets R_0 over $\{0, 1\}$ and for all positive rationals $r < 1$, $\{R | R$ is a regular expression and $L(R) \neq L(R_0)\}$ is not accepted by any n^r tape-bounded nondeterministic TM.*

We conclude by noting that the many parallels illustrated in this paper between the LBA problem and the "DPTIME = NDPTIME" questions, and between the complexity of predicates on the regular expressions and DFA deserve further study. The similarity between Theorems 18 and 27 is surprising as is the descriptive power of both the regular expressions and the DFA. In truth there is much more to the regular sets, DFA, regular expressions, etc., than one would have thought.

Bibliography

1. R. V. Book, *Comparing complexity classes,* Technical Report No. 1–73, Harvard University, Cambridge, Mass.

2. T. P. Baker, *Computational complexity and nondeterminism in flowchart programs,* Ph.D. Dissertation, Cornell University, Ithaca, N. Y., 1973.

3. R. Fagin, *Generalized first-order spectra and polynomial-time recognizable sets,* SIAM-AMS Proc., vol. 7, Amer. Math. Soc., Providence, R. I., 1974, pp. 43–73.

4. S. Greibach, *Jump PDA's, deterministic context-free languages, principal AFDL's and polynomial time recognition,* Proc. Fifth Annual ACM Sympos. on Theory of Computing, 1973, pp. 20–28.

5. J. Hartmanis, *On nondeterminacy in simple computing devices,* Acta Informat. 1 (1972), 334–336. MR 47 #6129.

6. H. B. Hunt, III, *On time and tape complexity of languages,* Ph.D. Dissertation, Cornell University, Ithaca, N. Y., 1973.

7. ———, *On time and tape complexity of languages.* I, Proc. Fifth Annual ACM Sympos. on Theory of Computing, 1973, pp. 10–19.

8. R. Karp, *Reducibilities among combinatorial problems,* R. Miller and J. Thatcher (eds.), Complexity of Computer Computations, Plenum Press, New York, 1972, pp. 85–104.

9. S.-Y. Kuroda, *Classes of languages and linear-bounded automata,* Information and Control 7 (1964), 207–223. MR 29 #6968.

10. P. S. Landweber, *Three theorems on phrase structure grammars of type* 1, Information and Control 6 (1963), 131–136. MR 29 #3291.

11. P. M. Lewis, R. E. Stearns and J. Hartmanis, *Memory bounds for recognition of context-free and context-sensitive languages,* IEEE Conference Record on Switching Circuit Theory and Logical Design, 1965, pp. 191–202.

12. A. Meyer and L. Stockmeyer, *The equivalence problem for regular expressions with squaring requires exponential space,* Conf. Record IEEE Thirteenth Annual Sympos. on Switching and Automata Theory, 1972, pp. 125–129.

13. J. Myhill, *Linearly bounded automata,* Wadd Technical Note 60–165, June 1960.

14. W. J. Savitch, *Relationships between nondeterministic and deterministic tape complexities,* J. Comput. System Sci. 4 (1970), 177–192. MR 42 #1605.

15. ———, *A note on multihead automata and context-sensitive languages,* Acta Informat. 2 (1973), 249–252.

16. J. C. Warkentin, and P. C. Fischer, *Predecessor machines and regression functions,* Proc. Fourth Annual ACM Sympos. on Theory of Computing, 1972, pp. 81–87.

17. J. E. Hopcroft, *On the equivalence and containment problems for context-free languages,* Math. Systems Theory 3 (1969), 119–124. MR 40 #2472.

18. H. B. Hunt III and D. J. Rosenkrantz, *Computational parallels between the regular and context-free languages,* Proc. Sixth Annual ACM Sympos. on Theory of Computing, 1974.

19. N. D. Jones, *Preliminary report: Reducibility among combinatorial problems in log n space,* Proc. the Seventh Annual Princeton Conference on Information Sciences and Systems, 1973, pp. 547–551.

20. L. J. Stockmeyer and A. R. Meyer, *Word problems requiring exponential time: Preliminary report,* Proc. Fifth Annual ACM Sympos. on Theory of Computing, 1973, pp. 1–9.

CORNELL UNIVERSITY

SIAM–AMS Proceedings
Volume 7
1974

Super-Exponential Complexity Of Presburger Arithmetic*

Michael J. Fischer and Michael O. Rabin

Abstract. Lower bounds are established on the computational complexity of the decision problem and on the inherent lengths of proofs for two classical decidable theories of logic: the first-order theory of the real numbers under addition, and Presburger arithmetic–the first-order theory of addition on the natural numbers. There is a fixed constant $c > 0$ such that for every (nondeterministic) decision procedure for determining the truth of sentences of real addition and for all sufficiently large n, there is a sentence of length n for which the decision procedure runs for more than 2^{cn} steps. In the case of Presburger arithmetic, the corresponding bound is $2^{2^{cn}}$. These bounds apply also to the minimal lengths of proofs for any complete axiomatization in which the axioms are easily recognized.

1. Introduction and main theorems. We present some results obtained in the Fall of 1972 on the computational complexity of the decision problem for certain theories of addition. In particular we prove the following results.

Let L be the set of formulas of the first-order functional (predicate) calculus written using just $+$ and $=$. Thus, for example, $\sim [x + y = y + z] \vee x + x = x$ is a formula of L, and $\forall x \exists y [x + y = y]$ is a sentence of L. Even though this is not essential, we shall sometimes permit the use of the

---

AMS (MOS) subject classifications (1970). Primary 02B10, 02G05, 68A20, 68A40.

* This research was supported in part by the National Science Foundation under research grant GJ–34671 to MIT Project MAC, and in part by the Artificial Intelligence Laboratory, an MIT research program sponsored by the Advanced Research Projects Agency, Department of Defense, under Office of Naval Research contract number N00014–70–A–0362–0003. The preparation of the manuscript was supported by the Hebrew University and the University of Toronto.

individual constants 0 and 1 in writing formulas of L. We assume a finite alphabet for expressing formulas of L, so a variable in general is not a single atomic symbol but is encoded by a sequence of basic symbols.

Let $\mathcal{N} = \langle N, + \rangle$ be the structure consisting of the set $N = \{0, 1, 2, \cdots\}$ of natural numbers with the operation $+$ of addition. Let $\mathrm{Th}(\mathcal{N})$ be the first-order theory of $\mathcal{N}$, i.e., the set of all sentences of L which are true in $\mathcal{N}$. For example, $\forall x \forall y[x + y = y + x]$ is in $\mathrm{Th}(\mathcal{N})$. Presburger has shown that $\mathrm{Th}(\mathcal{N})$ is decidable [6]. For brevity's sake, we shall call $\mathrm{Th}(\mathcal{N})$ *Presburger arithmetic* and denote it by PA.

THEOREM 1. *There exists a constant* $c > 0$ *such that for every decision procedure (algorithm)* AL *for* PA, *there exists an integer* n_0 *so that for every* $n > n_0$ *there exists a sentence* F *of* L *of length* n *for which* AL *requires more than* $2^{2^{cn}}$ *computational steps to decide whether* $F \in PA$.

The previous theorem applies also in the case of nondeterministic algorithms. This implies that not only algorithms require a super-exponential number of computational steps, but also proofs of true statements concerning addition of natural numbers are super-exponentially long. Let AX be a system of axioms in the language L (or in an extension of L) such that a sentence $F \in L$ is provable from AX ($\mathrm{AX} \vdash F$) if and only if $F \in \mathrm{PA}$. Let AX satisfy the condition that to decide for a sentence F whether $F \in \mathrm{AX}$, i.e., whether F is an axiom, requires a number of computational steps which is polynomial in the length $|F|$ of F.

THEOREM 2. *There exists a constant* $c > 0$ *so that for every axiomatization* AX *of Presburger arithmetic with the above properties there exists an integer* n_0 *so that for every* $n > n_0$ *there exists a sentence* $F \in PA$ *such that the shortest proof of* F *from the axioms* AX *is longer than* $2^{2^{cn}}$. *By the length of a proof we mean the number of its symbols.*

With slight modifications, Theorem 2 holds for any (consistent) system AX of axioms in a language M in which the notion of integer and the operation $+$ on integers are definable by appropriate formulas so that under this interpretation, all the sentences of PA are provable from AX. The ordinary axioms ZF for set theory have this property.

The result concerning super-exponential length of proof applies, in this more general case, to the sentences of M which are encodings of sentences of PA under the interpretation, i.e., to sentences which express elementary properties of addition of natural numbers.

The previous results necessarily involve a cut-point $n_0(\mathrm{AL})$ or $n_0(\mathrm{AX})$ at which the super-exponential length of computation or proofs sets in. It is significant that a close examination of our proofs reveals that $n_0(\mathrm{AL}) = O(|\mathrm{AL}|)$ and

$n_0(\mathrm{AX}) = O(|\mathrm{AX}|)$. Thus computations and proofs become very long quite early in the game.

The theory PA of addition of natural numbers is one of the simplest most basic imaginable mathematical theories. Unlike the theory of addition and multiplication of natural numbers, PA is decidable. Yet any decision procedure for PA is inherently difficult.

Let us now consider the structure $\mathcal{R} = \langle R, + \rangle$ of all *real* numbers R with addition. The theory $\mathrm{Th}(\mathcal{R})$ (in the same language L) is also decidable. In fact, to find a decision procedure for $\mathrm{Th}(\mathcal{R})$ is even simpler than a procedure for PA; this is mainly because $\mathcal{R}$ is a divisible group without torsion. Yet the following holds.

THEOREM 3. *There exists a constant* $d > 0$ *so that for the theory* $\mathrm{Th}(\mathcal{R})$ *of addition of real numbers, the statement of Theorem* 1 *holds with the lower bound* 2^{dn}.

Similarly for the length of proofs of sentences in $\mathrm{Th}(\mathcal{R})$.

THEOREM 4. *There exists a constant* $d > 0$ *so that for every axiomatization AX for* $\mathrm{Th}(\mathcal{R})$ *the statement of Theorem* 2 *holds with the lower bound* 2^{dn}.

COROLLARY 5. *The theory of addition and multiplication of reals* (*Tarski's algebra* [10]) *is exponentially complex in the sense of Theorems* 3 *and* 4.[1]

Ferrante and Rackoff [2] strengthen results of Oppen [5] to obtain decision procedures for $\mathrm{Th}(\mathcal{R})$ and PA which run in deterministic space 2^{cn} and $2^{2^{dn}}$ (and hence in deterministic time $O(2^{2^{cn}})$ and $O(2^{2^{2^{dn}}})$), respectively, for certain constants c and d. That time $2^{2^{cn}}$ is sufficient even for Tarski's algebra has been announced by Collins [1] and also by Solovay [8] who extends a result of Monk [4]. Any substantial improvement in our lower bounds would settle some open questions on the relation between time and space. For example, a lower bound of time 2^{n^2} for the decision problem for $\mathrm{Th}(\mathcal{R})$ would give an example of a problem solvable in space $S(n) = 2^{cn}$ but not in time bounded by a polynomial in $S(n)$ (cf. [9]).

Variations of the methods employed in the proofs of Theorems 1–4 lead to complexity results for the (decidable) theories of multiplication of natural numbers, finite Abelian groups, and other classes of Abelian groups. Some of these results are stated in §7 and will be presented in full in a subsequent paper.

The fact that decision and proof procedures for such simple theories are exponentially complex is of significance to the program of theorem proving by machine on the one hand, and to the more general issue of what is knowable in mathematics on the other hand.

---

[1] This result was obtained independently by V. Strassen.

2. Algorithms. Since we intend to prove results concerning the complexity of algorithms, we must say what notion of algorithm we use. Actually our methods of proof and our results are strong enough to apply to any reasonable class of algorithms or computing machines. However, for the sake of definiteness, we shall assume throughout this paper that our algorithms are the programs for Turing machines on the alphabet $\{0, 1\}$.

We proceed to give an informal description of these algorithms. The machine-tape is assumed to be one-way infinite extending to the right from an initial left-most square. At any given time during the progress of a computation, all but a finite number of the squares of the tape contain 0. An *instruction* has the form:

"i: If 0 then print X_0, move M_0, go to one of $i_1, i_2, \cdots$;
if 1 then print X_1, move M_1, go to one of $j_1, j_2, \cdots$."

Here $i, i_1, i_2, \cdots, j_1, j_2, \cdots$ are natural numbers, the so-called *instruction numbers*; X_0 and X_1 are either 0 or 1; and M_0 and M_1 are either R or L (for "move right" and "move left," respectively).

The possibility of going to one of several alternative instructions embodies the nondeterministic character of our algorithms. Another type of instruction is:

"i: *Stop.*"

Instructions are abbreviated by dropping the verbal parts. Thus, "3: 0, 1, L, 72, 5; 1, 1, R, 15, 3." is an example of an instruction. A program AL is a sequence $I_1, \cdots, I_n$ of instructions. For the sake of definiteness we assume that the instruction number of I_i is i and that I_n is the instruction "n: *Stop.*" Furthermore AL is assumed to be coded in the binary alphabet $\{0, 1\}$ in such a way that "*Stop*" also serves as an end-word indicating the end of the binary word AL.

Let $x \in \{0, 1\}^*$ be an input word. To describe the possible computations by the algorithm AL on x, we assume that x is placed in the left-most positions of the machine's tape and the scanning head is positioned on the left-most square of the tape. The computation starts with the first instruction I_1. A *halting computation* on x is a sequence $C = (I_{i_1}, \cdots, I_{i_m})$ of instructions of AL so that $i_1 = 1$ and $i_m = n$. At each step $1 \leqslant p \leqslant m$, the motion of the scanning head, the printing on the scanned square, and the transfer to the next instruction $I_{i_{p+1}}$ are according to the current instruction I_{i_p}. The *length* $l(C)$ of C is, by definition, m.

It is clear that a truly nondeterministic program may have several possible computations on a given input x.

3. Method for complexity proofs. Having settled on a definite notion of algorithm, we shall describe a general method for establishing lower bounds for theories of addition which are formalized in L. We do not develop our methods of proof in their fullest generality but rather utilize the fact that we deal with natural or real numbers to present the proofs in a more readily understandable and concrete form. The refinements and generalizations which are needed for other theories of addition will be introduced in a subsequent paper.

THEOREM 6. *Let $f(n)$ be one of the two functions 2^n or 2^{2^n}. Assume for a complete theory T that there exists a polynomial $p(n)$ and a constant $d > 0$ so that for every program AL and binary word x, there exists a sentence $F_{AL,x}$ with the following properties:*

(a) *$F_{AL,x} \in T$ if and only if some halting computation C of AL on x satisfies $l(C) \leqslant f(|x|)$.*

(b) $|F_{AL,x}| \leqslant d \cdot (|AL| + |x|)$.

(c) *$\sim F_{AL,x}$ is Turing machine calculable from AL and x in time less than $p(|AL| + |x|)$.*

(We recall that all our objects such as F, AL, etc. are binary words, and that $|w|$ denotes the length of w.)

Under these conditions, there exists a constant $c > 0$ so that for every decision algorithm AL for T there exists a number $n_0 = n_0(AL)$ so that for every $n > n_0$ there exists a sentence $\sigma \in T$ such that $|\sigma| = n$ and every computation by AL for deciding σ takes more than $f(cn)$ steps. Furthermore $n_0(AL) = O(|AL|)$.

PROOF. There exists a number $c > 0$ and an m_0 so that for $m \geqslant m_0$ we have

$$p(2m) + f(c \cdot (2dm + 1)) \leqslant f(m). \tag{1}$$

Namely, let $c < 1/(2d)$ and recall that $p(n)$ is a polynomial, whereas $f(n)$ is 2^n or 2^{2^n}.

Let AL be a (nondeterministic) decision algorithm for T. We construct a new algorithm AL_0 as follows. We do not care how AL_0 behaves on an input word x which is not a program. If x is a program, then AL_0 starts by constructing the sentence $F = \sim F_{x,x}$. The program AL_0 then switches to AL which works on the input F. If AL stops on F and determines that $F \in T$, then AL_0 halts; in all other cases AL_0 does not halt. Thus, for a program x as input, AL_0 halts if and only if the program x does *not* halt on the input x in fewer than $f(|x|)$ steps. Note that by possibly padding AL_0 with irrelevant instructions, we may assume that $m_0 \leqslant |\mathrm{AL}_0| \leqslant |\mathrm{AL}| + k$, where k is independent of AL.

Denote the binary word AL_0 by z and let σ be the sentence $\sim F_{z,z}$. $F_{z,z}$ cannot be true, for if it were true, then $\sim F_{z,z}$ would be false and AL_0

would not halt on z, whereas the truth of $F_{z,z}$ implies that z $(= AL_0)$ does halt on the input z (even in at most $f(|z|)$ steps), a contradiction.

Thus, σ is true and hence AL_0 $(= z)$ halts on z. The truth of σ also implies that every halting computation of AL_0 on z is longer than $f(|z|)$.

Let $m = |z|$. By (b), we have

$$n = |\sigma| \leq 2dm + 1. \tag{2}$$

Let t be the least number of steps that AL takes, by some halting computation, to decide σ. By the definition of AL_0 and the fact that fewer than $p(2m)$ steps are required to find $\sigma = \sim F_{z,z}$ from z (this follows from (c) and $|z| = m$), there is a halting computation of the program AL_0 on z requiring fewer than $p(2m) + t$ steps. By the truth of σ, $p(2m) + t > f(m)$. Using (1) and (2), $t > f(c \cdot (2dm + 1)) \geq f(cn)$.

Take n_0 to be $n = |\sigma|$. Then $n_0 \leq 2dm + 1 \leq 2d(|AL| + k) + 1$, so $n_0 = O(|AL|)$. The fact that the result holds for AL and every $n > n_0$ (with possibly a smaller constant c) is obtained by first padding AL_0 by irrelevant instructions, and then padding the resulting σ by prefixing a quantifier $\exists x_j$ of an appropriate length, where $|\exists x_j| = 1 + |j|$. The details are left to the reader. □

For utilizing Theorem 6 we need a method for constructing sentences $F_{AL,w}$ with the properties (a)–(c). One such method is provided by

THEOREM 7. *Let* $\mathfrak{A} = \langle A, + \rangle$ *be an additive structure such that* $N \subseteq A$, *and on* N *the operation* $+$ *is ordinary addition. Let* $f(n)$ *again be one of the functions* 2^n *or* 2^{2^n}. *Assume that* $T = \mathrm{Th}(\mathfrak{A})$ *is a theory of addition (formalized in the language L) for which there exists* $c > 0$ *such that, for every* n *and for every binary word* w, $|w| = n$, *there exist formulas* $I_n(y)$, $J_n(y)$, $S_n(x, y)$ *and* $H_w(x)$ *with the following properties*:

(i) $|S_n(x, y)| \leq cn$, $|I_n(y)| \leq cn$, $|J_n(y)| \leq cn$, *and* $|H_w(x)| \leq cn$.

(ii) $I_n(b)$ *is true in* $\mathfrak{A}$ *for* $b \in A$ *if and only if* $b \in N$ *and* $b < f(n)^2$. $J_n(b)$ *is true exactly for* $b = f(n)$.

(iii) S_n *codes all binary sequences of length* $f(n)^2$. *Namely, for every binary sequence* $\beta \in \{0, 1\}^*$, $|\beta| = f(n)^2$, *there exists an* $\alpha \in A$ *so that, for* $i \in N$, $0 \leq i < f(n)^2$, $S_n(\alpha, i)$ *is true in* $\mathfrak{A}$ *if* $\beta(i) = 1$, *and* $S_n(\alpha, i)$ *is false if* $\beta(i) = 0$, *where, for any sequence* β, $\beta(i)$ *denotes the* $(i + 1)$*st element of* β, $0 \leq i < |\beta|$.

(iv) $H_w(x)$ *is true for* $\alpha \in A$ *if and only if the first* $f(n)$ *symbols of the sequence coded by* α *in the sense of* (iii) *have the form* $w0^p$, $p = f(n) - |w|$.

(v) $S_n(x, y)$, $I_n(y)$, $J_n(y)$ *and* $H_w(x)$ *are Turing machine calculable from* n *and* w *in a polynomial number of steps.*

From such formulas S_n, I_n, J_n *and* H_w, *a formula* $F_{AL,w}$ *with the properties* (a)–(c) *can be constructed, so that* T *satisfies the conclusion of Theorem* 6.

PROOF. We shall describe, by use of sequences of length $f(n)^2$, all possible halting computations of length at most $f(n)$ of a program AL on an input w. Let $C = (I_{i_1}, \cdots, I_{i_m})$ be such a computation. Assume that AL has k instructions; by our notational conventions every computation starts with the first instruction I_1 and the last instruction I_k of AL is: "k: *Stop*." Thus, in C, $i_1 = 1$ and $i_m = k$.

Let us adopt the convention that after the stop instruction, the scanning head, the (stop) instruction, and the tape contents stay stationary and unchanged at all subsequent time instants. Since $m \leqslant f(n)$, the scanning head never moves beyond $f(n)$ squares from the initial left-most square of the tape. We assume also that the Turing machine never attempts to shift its head left off of the beginning of the tape.

The progress of the computation C on the input w will be described by stringing together $f(n)$ instantaneous descriptions of the computation in the following manner. Let W_j be the first (left-most) $f(n)$ symbols of the tape at time j, $1 \leqslant j \leqslant f(n)$. Then the string $W_1 W_2 \cdots W_{f(n)} = W \in \{0, 1\}^*$ codes all the relevant information concerning the tape contents during the computation C. We have $|W| = f(n)^2$. Also, $W_m = W_{m+1} = \cdots$.

To trace the motion of the scanning head and the sequence of instructions during the computation C, we define $U_j \in \{0, 1, \cdots, k\}^*$ to be $0^{p_j}\, i_j\, 0^{q_j}$ where $p_j + q_j + 1 = f(n)$ and p_j is the distance at time j of the scanning head from the start square, $1 \leqslant j \leqslant f(n)$. Recall that i_j is the instruction number of the jth instruction executed in C. Also $i_m = i_{m+1} = \cdots = k$, the stop instruction. Put $U = U_1 U_2 \cdots U_{f(n)}$. We have $|U| = f(n)^2$.

The fact that the pair (W, U), where $W \in \{0, 1\}^*$, $U \in \{0, 1, \cdots, k\}^*$, $|W| = |U| = f(n)^2$, describes a halting computation of AL on w, is equivalent to a number of statements which say, roughly, that the first $f(n)$ symbols are the initial configuration, that the transformation from a block of $f(n)$ symbols to the next block is by an instruction of AL, and that U contains k (the number of the halting instruction). More precisely, (W, U) codes a halting computation of length at most $f(n)$ of AL on w, where $|w| = n$, if and only if the following hold:

(α) $W(0) \cdots W(f(n) - 1) = w0^p$, $p = f(n) - |w|$.

(β) $U(0) \cdots U(f(n) - 1) = 10^{f(n)-1}$.

(γ) If $U(i) = 0$ and $i + f(n) < f(n)^2$, then $W(i + f(n)) = W(i)$.

(δ) If $U(i) = q$, $i + f(n) + 1 < f(n)^2$, $0 < q < k$, $W(i) = 0$, and I_q is, say, "q: 0, 1, R, $k_1, \cdots, k_t$; 1, $\cdots$," then $W(i + f(n)) = 1$, $U(i + f(n) + 1) = k_1$ or $U(i + f(n) + 1) = k_2$ or etc. (similarly for other instruction and tape-symbol combinations).

(ϵ) If, for $f(n) < i < f(n)^2$, $U(i) \neq 0$, then exactly one of $U(i - f(n)) \neq 0$, or $U(i - f(n) - 1) \neq 0$, or $U(i - f(n) + 1) \neq 0$ holds. Also, if $U(i) \neq 0$, then $U(i \pm 1) = U(i \pm 2) = 0$ (if they are defined).

(ζ) $U(i) = k$ for some i, $0 \leqslant i < f(n)^2$. If $U(i) = k$ and $i + f(n) < f(n)^2$, then $U(i + f(n)) = k$ and $W(i + f(n)) = W(i)$.

From the assumption that (W, U) satisfies (α)–(ζ), it can be proved by induction on $1 \leqslant j < f(n)$ that (W_{j+1}, U_{j+1}) is an instantaneous description which follows from (W_j, U_j) by an application of the instruction I_{i_j} whose number appears in U_j. Also, $(W_{f(n)}, U_{f(n)})$ is a halting instantaneous description.

Thus, the existence of a pair (W, U), $W \in \{0, 1\}^*$, $U \in \{0, 1, \cdots, k\}^*$, $|W| = |U| = f(n)^2$, which satisfies (α)–(ζ) is a necessary and sufficient condition for the existence of a halting computation C on w with $l(C) \leqslant f(n)$.

Conditions (i)–(v) provide means for making statements about arbitrary (0, 1) sequences of length $f(n)^2$, about integers $0 \leqslant i < f(n)^2$, and about the integer $f(n)$, all by use of formulas of L of size $O(n)$. Also, the ordinary ordering $\leqslant$ on N restricted to integers of size less than $f(n)^2$ can be expressed by the length $O(n)$ formula

$$x \leqslant_n y \leftrightarrow \exists z[I_n(x) \wedge I_n(y) \wedge I_n(z) \wedge x + z = y].$$

Hence, the existence of (W, U) satisfying (α)–(ζ) can be expressed by a sentence $F_{\mathrm{AL},w} = F$ with the desired properties (a)–(c). Namely, express 0, 1, $\cdots$, k in binary notation by words of equal length $p = |k|$. Then, via $S_n(x, y)$, a single element $a \in A$ exists which codes W, and elements $a_1, \cdots, a_p \in A$ code U. The sentence F will start with quantifiers and relativization:

$$F = \exists x \exists x_1 \cdots \exists x_p \forall y \forall z [I_n(y) \wedge J_n(z) \rightarrow E_\alpha \wedge E_\beta \wedge E_\gamma \wedge E_\delta \wedge E_\epsilon \wedge E_\zeta].$$

x codes the sequence W and $x_1, \cdots, x_p$ together code the sequence U. The clauses $E_\alpha \cdots E_\zeta$ express the corresponding conditions (α)–(ζ). Thus, for example, E_α is $H_w(x)$; E_β is $H_{u_1}(x_1) \wedge H_{u_2}(x_2) \wedge \cdots \wedge H_{u_p}(x_p)$, where $u_1 = 10^{n-1}$ and $u_j = 0^n$, $2 \leqslant j \leqslant p$; and E_γ is

$$[\sim S_n(x_1, y) \wedge \cdots \wedge \sim S_n(x_p, y) \wedge I_n(y + z) \\ \rightarrow [S_n(x, y + z) \leftrightarrow S_n(x, y)]].$$

The reader can supply the details of the construction of the remaining expressions E_δ, E_ϵ and E_ζ and verify that, altogether, the $F_{AL,w}$ thus formed satisfies (a)–(c) of Theorem 6. □

4. Proof of Theorem 3 (real addition). We start by showing that for the theory $\mathrm{Th}(\mathcal{R})$ of real addition, there exist formulas $S_n(x, y)$, $I_n(y)$, etc. as postulated in Theorem 7 with $f(n) = 2^n$, thereby proving Theorem 3. Several of the results in this section will play a role later on in the proof for PA.

Let $F(x, y)$ be any formula and consider the conjunction

$$G = F(x_1, y_1) \wedge F(x_2, y_2) \wedge F(x_3, y_3).$$

It is readily seen that $G \leftrightarrow G_1$ where

$$G_1 = \forall x \forall y [((x = x_1 \wedge y = y_1) \vee (x = x_2 \wedge y = y_2) \vee (x = x_3 \wedge y = y_3)) \rightarrow F(x, y)].$$

Note that $|G| = 3 \cdot |F(x, y)|$, whereas $|G_1| = |F(x, y)| + c$, where c is independent of $F(x, y)$. A similar rewriting exists for formulas F with more than two variables and for conjunctions of more than three instances of F. The above device, discovered independently by several people including V. Strassen, is a special case of a more general theorem due to M. Fischer and A. Meyer.

THEOREM 8. *There exists a constant* $c > 0$ *so that for every* n *there is a formula* $M_n(x, y, z)$ *of* L *such that, for real numbers* A, B, C,

$$M_n(A, B, C) \text{ is true} \leftrightarrow A \in N \wedge A < 2^{2^n} \wedge AB = C.$$

Also, $|M_n(x, y, z)| \leq c(n + 1)$ *and* $M_n(x, y, z)$ *is Turing machine computable from* n *in time polynomial in* n.

PROOF. The construction of $M_n(x, y, z)$ will be inductive on n. For $n = 0$ we have $2^{2^0} = 2$ and we define $M_0(x, y, z)$ as $[x = 0 \wedge z = 0] \vee [x = 1 \wedge z = y]$.

From M_k we get M_{k+1} by observing that $x \in N$ and $x < 2^{2^{k+1}}$ if and only if there exist $x_1, x_2, x_3, x_4 \in N$ all less than 2^{2^k} so that $x = x_1x_2 + x_3 + x_4$. For this decomposition we have $z = xy = x_1(x_2y) + x_3y + x_4y$. Hence, $M_{k+1}(x, y, z)$ is equivalent to

$$\exists u_1 u_2 \cdots u_5 x_1 \cdots x_4 [M_k(x_1, x_2, u_1) \wedge M_k(x_2, y, u_2) \wedge M_k(x_1, u_2, u_3) \wedge M_k(x_3, y, u_4) \wedge M_k(x_4, y, u_5) \wedge x = u_1 + x_3 + x_4 \wedge z = u_3 + u_4 + u_5].$$

(Strictly speaking, a triple sum such as $u_1 + x_3 + x_4$ should be written as a chain of sums of two variables, but we shall not do it here.) Now, $|M_{k+1}| \geqslant 5|M_k|$, which will not do. However, by using the device preceding the theorem, the five occurrences of M_k can be replaced by a single occurrence to yield M_{k+1}. Thus, $|M_{k+1}(x, y, z)| \leqslant |M_k(x, y, z)| + c$ for an appropriate $c > 0$. Hence, $|M_n(x, y, z)| \leqslant c(n + 1)$. (We assume c is chosen large enough so $c \geqslant |M_0(x, y, z)|$.)

Actually, for the above bound to hold, it is necessary to show that the number of distinct variable names in M_n does not grow with n, for to encode one of v variables requires (on the average) a string of length $O(\log v)$. In fact, 15 different variable names are sufficient to express M_n. This is because the new variables introduced in constructing M_{k+1} from M_k need only be distinct from each other and from the *free* variables of M_k; however no difficulty arises if they coincide with variables *bound* inside M_k. A closer look at the construction of M_{k+1} shows that 12 new variables are introduced, which must be distinct from the three free variables of M_k, giving a total of 15 distinct names needed. □

COROLLARY 9. *The formula* $M_n(x, 0, 0)$ *is true for a real number* x *if and only if* $x \in N$ *and* $x < 2^{2^n}$.

The natural numbers $x < 2^{2^n}$ code all binary sequences of length 2^n. Namely, write x in binary notation

$$x = x(0) + x(1) \cdot 2 + \cdots + x(2^n - 1) \cdot 2^{2^n - 1}.$$

We use the function 2^i to obtain the element $x(i)$ of x.

THEOREM 10. *There exists a formula* $\mathrm{Pow}_n(x, y, z)$ *such that, for integers* a, b, c *for which* $0 \leqslant a, b^a, c < 2^{2^n}$, $\mathrm{Pow}_n(a, b, c)$ *is true if and only if* $b^a = c$. *Also,* $|\mathrm{Pow}_n(x, y, z)| \leqslant d(n + 1)$ *for an appropriate* $d > 0$ *and all* n.

PROOF. Construct, by induction on k, a sequence $E_k(x, y, z, u, v, w)$ of formulas with the property that for integers a, b, c for which $0 \leqslant a < 2^{2^k}$, $0 \leqslant b^a, c < 2^{2^n}$ and real numbers A, B, C, $E_k(a, b, c, A, B, C)$ is true in $\langle R, +\rangle$ if and only if $A \in N$, $A < 2^{2^n}$, $b^a = c$, and $AB = C$. Thus, E_k has M_n built into it since

$$E_k(0, 1, 1, A, B, C) \leftrightarrow M_n(A, B, C).$$

The case $k = 0$ is given by

$$[(x = 0 \wedge z = 1) \vee (x = 1 \wedge z = y)] \wedge M_n(u, v, w).$$

To obtain $E_{k+1}(x, y, z, u, v, w)$ from E_k, we again use the decomposition $x = x_1x_2 + x_3 + x_4$ of every integer $0 \leqslant x < 2^{2^{k+1}}$ in terms of integers $0 \leqslant x_1, x_2, x_3, x_4 < 2^{2^k}$. Then we have $y^x = (y^{x_1})^{x_2} \cdot y^{x_3} \cdot y^{x_4}$. Now, y^{x_1} is expressed by a z_1 such that $E_k(x_1, y, z_1, 0, 0, 0)$; then $(y^{x_1})^{x_2}$ is a z_2 such that $E_k(x_2, z_1, z_2, 0, 0, 0)$, etc. Whenever we have to write a product such as x_1x_2 or $(y^{x_1})^{x_2} \cdot y^{x_3}$, we use the formula $E_k(0, 1, 1, u, v, w)$. In this way we can write the formula $E_{k+1}(x, y, z, u, v, w)$. Using the usual device of contracting a conjunction of instances of E_k into one occurrence, we see that $|E_{k+1}| \leqslant |E_k| + d$ for some $d > 0$, and hence $|E_n| \leqslant d(n + 1) + c(n + 1)$, where $c(n + 1)$ is the bound on the length of M_n. As before, only a bounded number of variable names are needed.

Recalling the definition of $E_k(x, y, z, u, v, w)$, we see that

$$\mathrm{Pow}_n(x, y, z) \leftrightarrow E_n(x, y, z, 0, 0, 0)$$

has the desired properties. □

THEOREM 11. *There exists a formula $S_n(x, y)$ of L which for $x, y \in R$ is true in $\langle R, +\rangle$ if and only if x and y are integers, $x < 2^{2^{2n}}$ and $y < 2^{2n}$, and the $(y + 1)$st digit $x(y)$ of x, counting from the low-order end of the binary representation of x, is* 1. *The formula $S_n(x, y)$ satisfies the conditions of Theorem 7 for $f(n) = 2^n$.*

PROOF. That x and y are integers in the appropriate ranges is easily expressible by formulas of size $O(n)$. Recall that for the integers which satisfy $M_{2n}(x, 0, 0)$, i.e., $0 \leqslant x < 2^{2^{2n}}$, the ordering $\leqslant$ is expressible by a formula of length $O(n)$.

Now $x(y) = 1$ if and only if there exists an integer z, $2^y \leqslant z < 2^{y+1}$ so that $x \geqslant z$ and 2^{y+1} divides $x - z$. This fact is easily expressible by a formula $S_n(x, y)$ of L using Pow_{2n} and M_{2n}.

That formulas $I_n(y)$ and $J_n(y)$ with the properties listed in Theorem 7 exist is immediate. Thus to finish the proof of Theorem 3 we need the following.

THEOREM 12. *For every binary word w, $|w| = n$, there exists a formula $H_w(x)$ of L which is true in $\langle R, +\rangle$ for an integer $0 \leqslant x < 2^{2^{2n}}$ if and only if $x(0) \cdots x(2^n - 1) = w0^p$, $p = 2^n - n$. The formula $H_w(x)$ satisfies the conditions of Theorem* 7.

PROOF. Define for binary words u, by induction on $|u|$, formulas $K_u(z)$ as follows.

$$K_0(z) \leftrightarrow z = 0,$$
$$K_1(z) \leftrightarrow z = 1,$$
$$K_{u0}(z) \leftrightarrow \exists y[K_u(y) \wedge z = y + y],$$
$$K_{u1}(z) \leftrightarrow \exists y[K_u(y) \wedge z = y + y + 1].$$

Clearly, if $K_w(z)$ is true, then, considered as a sequence, z satisfies $w(i) = z(i)$ for $0 \leqslant i < |w|$, $z(i) = 0$ for $i \geqslant |w|$. Using this $K_w(z)$ and the formulas $S_n(x, y)$ and $J_n(y)$, we can write the formula $H_w(x)$ by formally expressing the statement that for z such that $K_w(z)$, $x(i) = z(i)$, $0 \leqslant i < 2^n$. $\square$

Thus we have proved, for $\mathrm{Th}(\mathcal{R})$, the existence of formulas $S_n(x, y)$, $I_n(y)$, $J_n(y)$, and $H_w(x)$ which satisfy the conditions of Theorem 7 for $f(n) = 2^n$. This completes the proof of Theorem 3.

5. Proof of Theorem 4 (lengths of proofs for real addition). We now show that for $\mathrm{Th}(\mathcal{R})$ proofs are also exponentially long. This is an easy consequence of Theorem 3.

Let AX be a consistent system of axioms which is complete for $\mathrm{Th}(\mathcal{R})$, i.e., every sentence $F \in \mathrm{Th}(\mathcal{R})$ is provable from AX ($\mathrm{AX} \vdash F$). Furthermore, there exists an algorithm B which decides in polynomial time $p(|G|)$ for a sentence G of L whether $G \in \mathrm{AX}$.

Let c be the constant of Theorem 3. For every polynomial $q(x)$, there exists a constant $0 < d$ so that from a certain point on, $q(2^{dn}) < 2^{cn}$.

Construct a nondeterministic algorithm AL for $\mathrm{Th}(\mathcal{R})$ as follows. Given a sentence F, AL writes down (nondeterministically) a binary sequence P. Then AL checks whether P is a proof of F from AX or a proof of $\sim F$ from AX. The computation halts only if one of the two possibilities occurs. Because of the assumptions on AX, this check can be made in a polynomial number of steps $h(|P|)$. Thus the whole computation, if it halts, requires $|P| + h(|P|) = q(|P|)$ steps. If every true sentence F would have a proof P with $|P| < 2^{dn}$ where $n = |F|$, then for every such F there would be some halting computation of length less than $q(2^{dn})$, i.e., also less than 2^{cn} for all sufficiently large n, a contradiction.

6. Proof of Theorems 1 and 2 (Presburger arithmetic). The proof for Theorem 1 follows closely along the lines of the proof of Theorem 3 and utilizes our previous results. In particular we note that Theorems 8–10 apply, as they stand and with the same proofs, to PA. Note also that the order $\leqslant$ on N is definable in PA using $+$. Throughout this section, let $f(n)$ be 2^{2^n}.

THEOREM 13. *There exists a function* $g(n) \geqslant 2^{f(n)^2} = 2^{2^{2^{n+1}}}$ *so that for every* n *there exists a formula* $\mathrm{Prod}_n(x, y, z)$ *with the following properties. For integers* A, B, C,

$\text{Prod}_n(A, B, C)$ *is true in* $N \leftrightarrow A, B, C < g(n)$ *and* $AB = C$.

There exists a constant $c > 0$ *so that* $|\text{Prod}_n| \leqslant c(n+1)$ *for all* n. *The formula* Prod_n *is Turing machine constructible from* n *in time polynomial in* n.

PROOF. We shall use the Prime Number Theorem which says that the number of primes smaller than m is asymptotically equal to $m/\log_e m$; hence bigger than $m/\log_2 m$ for all sufficiently large m. Thus, for $m = 2^{2^{n+2}}$, the number of primes $p < m$ exceeds $2^{2^{n+2}}/2^{n+2} > 2^{2^{n+1}} = f(n)^2$. Let $g(n) = \Pi_{p<m}\, p$, where p runs over primes, $m = 2^{2^{n+2}}$; then $g(n) \geqslant 2^{f(n)^2}$ since $2 \leqslant p$ for all primes.

By use of the formula $M_{n+2}(x, y, z)$, we can write two formulas $\text{Res}_{n+2}(x, y, z)$ and $P_{n+2}(x)$ of length $O(n)$ with the following meanings. Let $\text{res}(x, y)$ denote the residue (remainder) of x when divided by y. Then

$$\text{Res}_{n+2}(x, y, z) \leftrightarrow [y < 2^{2^{n+2}} \wedge \text{res}(x, y) = z],$$

$$P_{n+2}(x) \leftrightarrow [x < 2^{2^{n+2}} \text{ and } x \text{ is prime}].$$

The formula Res_{n+2} is written in L as

$$z < y \wedge \exists q \exists w [M_{n+2}(y, q, w) \wedge x = w + z].$$

We recall that, for any q and w, $M_{n+2}(y, q, w)$ holds if and only if $y < 2^{2^{n+2}}$ and $yq = w$.

The formula $P_{n+2}(x)$ is, simply,

$$M_{n+2}(x, 0, 0) \wedge \forall y \forall z [M_{n+2}(y, z, x) \rightarrow [y = 1 \vee y = x]].$$

By formally saying that $x \geqslant 1$ is the smallest integer divisible by all primes $p < 2^{2^{n+2}}$, we can write a formula $G_{n+2}(x)$ which is true precisely for $x = g(n)$. Now $\text{Prod}_n(x, y, z)$ is true if and only if

$$x, y, z < g(n) \wedge \forall u [u < 2^{2^{n+2}} \rightarrow \text{res}(x, u) \cdot \text{res}(y, u) = \text{res}(z, u)]. \tag{3}$$

Namely, this implies that $xy = z (\text{mod } p)$ for all $p < 2^{2^{n+2}}$, which together with $x, y, z < g(n)$ is equivalent to $xy = z$. Now, by use of $G_{n+2}(x)$, $M_{n+2}(x, y, z)$ and $\text{Res}_{n+2}(x, y, z)$, the above relation (3) can be expressed by a formula Prod_n with the desired properties. □

Exponentiation can be defined just as in the proof of Theorem 10 except that we now use $\text{Prod}_n(x, y, z)$ instead of $M_n(x, y, z)$ to obtain a sequence of formulas $E'_k(x, y, z, u, v, w)$. For integers a, b, c, A, B, C for which $0 \leqslant a < 2^{2^k}$ and $0 \leqslant b^a, c < g(n)$, $E'_k(a, b, c, A, B, C)$ is true in N if and only if $A, B, C < g(n)$, $b^a = c$, and $AB = C$. Also $|E'_n| = O(n)$.

Having now multiplication up to $g(n)$ and exponentiation 2^i up to $i < 2^{2^{n+1}}$ expressed by formulas of length $O(n)$, we can code sequences of length $2^{2^{n+1}} = f(n)^2$ in exactly the same manner as in §4. This completes the proof of Theorem 1 by again appealing to Theorem 7.

The proof of Theorem 2 now follows exactly the lines of the proof of Theorem 4 given in §5.

7. Other results. The techniques presented in this paper for proving lower bounds on logical theories may be extended in a number of directions to yield several other results. We outline some of them below without proof; they will be presented in full in a subsequent paper.

THEOREM 14. *Let* $\mathfrak{A}$ *be any class of additive structures, so if* $\mathsf{A} = \langle A, +\rangle \in \mathfrak{A}$, *then* $+$ *is a binary associative operation on* A. *Let* $\mathrm{Th}(\mathfrak{A})$ *be the set of sentences of* L *valid in every structure of* $\mathfrak{A}$. *Assume* $\mathfrak{A}$ *has the property that, for every* $k \in N$, *there is a structure* $\mathsf{A}_k = \langle A_k, +\rangle \in \mathfrak{A}$ *and an element* $u \in A_k$ *such that the elements* $u, u + u, u + u + u, \cdots, k \cdot u$ *are distinct. Then the statement of Theorem* 1 *holds for* $\mathrm{Th}(\mathfrak{A})$ *with the lower bound* 2^{dn} *for some* $d > 0$.

Theorem 3 is an immediate corollary of this result, taking $\mathfrak{A}$ to be the class of just the one structure $\mathcal{R} = \langle R, +\rangle$. Some other classes to which the result applies are the following:

(1) the complex numbers under addition,

(2) finite cyclic groups,

(3) rings of characteristic p,

(4) finite Abelian groups,

(5) the natural numbers under multiplication.

The proof of Theorem 14 extends the ideas of §4. The element $n \cdot u$ is used as the representation of the integer n, and u itself is selected by existential quantification.

Special properties of certain theories permit us to obtain still larger lower bounds on the decision problem. For example, we get a lower bound of time $2^{2^{cn}}$ for (4), the theory of finite Abelian groups. This is obtained by encoding integers up to 2^{2^n} by formulas of length $O(n)$ just as in Theorem 14, but instead of representing a sequence by an integer, we let the structure itself encode the sequence. Let G be a finite Abelian group. Then the element $s(i)$ of the sequence s encoded by G is 1 if and only if G contains an element x of order p_i, where p_i is the $(i + 1)$st prime. The necessity of using primes as indices instead of integers considerably complicates the analog of Theorem 7.

Another example where we get still larger bounds is (5), the theory of multiplication of the natural numbers (MULT). That MULT is at least as hard as PA is immediate, for the powers of 2 under multiplication are isomorphic to N, and the property of being a power of 2 can be expressed in MULT (assuming we have the constant 2; otherwise we use an arbitrary prime). In fact, the bound can be increased yet another exponential to time $2^{2^{2^{cn}}}$ by using the encoding which associates a sequence s to a positive integer m, where $s(i) = 1$ if and only if q_i divides m, where q_i is the $(i + 1)$st prime in some fixed (but arbitrary) ordering of the primes. Again we are forced to use the primes as indices, and again the analog to Theorem 7 is considerably complicated. Rackoff [7] shows a corresponding upper bound of deterministic space $2^{2^{2^{dn}}}$.

Acknowledgement. The first examples of exponential and larger lower bounds on the complexity of logical theories and certain word problems from the theory of automata were obtained by A. R. Meyer [3] and L. J. Stockmeyer [9], and we gratefully acknowledge the influence of their ideas and techniques on our work. We are also indebted to C. Rackoff, R. Solovay, and V. Strassen for several helpful ideas and suggestions which led to and are incorporated in the present paper.

References

1. G. E. Collins, *Quantifier elimination for real closed fields by cylindrical algebraic decomposition. Preliminary report*, Proc. EUROSAM 74 Conf., Stockholm, 1974 (to appear).

2. J. Ferrante and C. Rackoff, *A decision procedure for the first order theory of real addition with order*, Project MAC Technical Memorandum 33, M. I. T., Cambridge, Mass., May 1973, 16 pp.

3. A. R. Meyer, *Weak monadic second order theory of successor is not elementary-recursive*, Project MAC Technical Memorandum 38, M. I. T., Cambridge, Mass., 1973, 24 pp.

4. L. Monk, *An elementary-recursive decision procedure for* Th(R, +, •), manuscript, Dept. of Math., Univ. of California at Berkeley, 1974, 16 pp.

5. D. C. Oppen, *Elementary bounds for Presburger arithmetic*, Proc. 5th ACM Symp. on Theory of Computing, 1973, 34–37.

6. M. Presburger, *Über die Vollständigkeit eines gewissen Systems der Arithmetic ganzer Zahlen in welchem die Addition als einzige Operation hervortritt*, Comptes-rendus du I Congrès des Mathématiciens des Pays Slaves, Warsaw, 1930, pp. 92–101, 395.

7. C. Rackoff, *Complexity of some logical theories*, Doctoral thesis, Dept. of Electrical Engineering, M. I. T., Cambridge, Mass., 1974, 129 pp.

8. R. Solovay, private communication.

9. L. J. Stockmeyer, *The complexity of decision problems in automata theory and logic*, Project MAC Technical Report 133, M. I. T., Cambridge, Mass., 1974, 224 pp.

10. A. Tarski, *A decision method for elementary algebra and geometry*, 2nd ed., Univ. of California Press, Berkeley and Los Angeles, Calif., 1951. MR 13, 423.

MASSACHUSETTS INSTITUTE OF TECHNOLOGY

HEBREW UNIVERSITY

SIAM–AMS Proceedings
Volume 7
1974

Generalized First-Order Spectra and Polynomial-Time Recognizable Sets[1]

Ronald Fagin

1. Introduction. A *finite structure* is a nonempty finite set, along with certain given functions and relations on the set. For example, a finite group is a set A, along with a binary function $\cdot: A \times A \longrightarrow A$. If σ is a sentence of first-order logic, then the *spectrum* of σ is the set of cardinalities of finite structures in which σ is true. For example, let σ be the following first-order sentence, where f is a "unary function symbol":

$$(1) \qquad \forall x(f(x) \neq x) \wedge \forall x \forall y(f(x) = y \leftrightarrow f(y) = x).$$

Then the spectrum of σ is the set of even positive integers. For, if σ is true about a finite structure $\mathfrak{A} = \langle A; g\rangle$, where A is the universe and g: $A \longrightarrow A$ (g is the "interpretation" of f), then $\mathfrak{A}$ must look like Figure 1, where $a \longrightarrow b$ means $g(a) = b$.

$$a_1 \longleftrightarrow a_2$$
$$a_3 \longleftrightarrow a_4$$
$$a_5 \longleftrightarrow a_6$$
$$\vdots$$

FIGURE 1

AMS (MOS) subject classifications (1970). Primary 02H05, 68A25; Secondary 02B10, 02E15, 02F10, 68A25, 94A30.

[1]This paper is based on a part of the author's doctoral dissertation in the Department of Mathematics at the University of California, Berkeley. Part of this work was carried out while the author was a National Science Foundation Graduate Fellow and was supported by NSF grant GP-24352.

So, the finite structure $\mathfrak{A}$ has even cardinality. And conversely, for each even positive integer n, there is a way to impose a function on n points to make σ be true about the resulting finite structure.

As a more interesting example, let σ be the conjunction of the field axioms–for example, one conjunct of σ is

$$\forall x \forall y \forall z (x \cdot (y + z) = x \cdot y + x \cdot z).$$

Then the spectrum of σ is the set of powers of primes.

In 1952, H. Scholz [21] posed the problem of characterizing spectra, that is, those sets (of positive integers) which are the spectrum of a sentence of first-order logic. It is well known that every spectrum is recursive: For, assume that we are given a first-order sentence σ and a positive integer n. To determine if n is in the spectrum of σ, we simply systematically write down all finite structures (up to isomorphism) of cardinality n of the relevant type, and test them one by one to see if σ is true in any of them. It is also well known that not every recursive set is a spectrum: We simply form the diagonal set D such that $n \in D$ iff n is not in the nth spectrum (the details are easy to work out).

In 1955, G. Asser [1] posed the problem of whether or not the complement of every spectrum is a spectrum. For example, it is not immediately clear how to write a first-order sentence with spectrum the numbers which are *not* powers of primes.

Note that the spectrum of the sentence (1) is the set of positive integers n for which the following so-called "existential second-order sentence" is true about some (each) set of n points:

$$\exists f (\forall x (f(x) \neq x) \wedge \forall x \forall y (f(x) = y \leftrightarrow f(y) = x)).$$

This suggests a generalization, which is due to Tarski [23]. Let σ be an existential second-order sentence (we will define this and other concepts precisely later), which may have not only bound but free predicate (relation) and function variables. Then the *generalized spectrum* of σ is the class of structures (not numbers) for which σ is true. Let us give some examples. The first few examples will deal with finite structures with a single binary realtion. We can think of these as finite directed graphs.

1. *The class of all k-colorable finite directed graphs, for fixed* $k \geqslant 2$. A (directed) graph $\mathfrak{A} = \langle A; G \rangle$ is *k-colorable* if the universe A of $\mathfrak{A}$ can be partitioned into k subsets $A_1, \cdots, A_k$ such that $\sim Gab$ holds if a and b are in the same subset of the partition. This class is a generalized spectrum, via the following existential second-order sentence, in which Q is a binary predicate symbol which represents the graph relation, and $C_1, \cdots, C_k$ are unary predicate

symbols ($\bigwedge_{i=1}^{k} \phi_i$ abbreviates $\phi_1 \wedge \cdots \wedge \phi_k$; similarly for $\bigvee_{i=1}^{k} \phi_i$):

$$\exists C_1 \cdots \exists C_k \left(\forall x \left(\bigvee_{i=1}^{k} C_i x \right) \wedge \forall x \left(\bigwedge_{i \neq j} \sim (C_i x \wedge C_j x) \right) \right.$$

$$\left. \wedge \forall x \forall y \left(Qxy \rightarrow \bigwedge_{i=1}^{k} \sim (C_i x \wedge C_i y) \right) \right).$$

2. *The class of finite directed graphs with a nontrivial automorphism.* This class is a generalized spectrum, via the following existential second-order sentence, in which Q is as before, and f is a unary function symbol:

$$\exists f (\exists x (f(x) \neq x) \wedge \forall x \forall y (f(x) = f(y) \rightarrow x = y)$$

$$\wedge \forall x \forall y (Qxy \leftrightarrow Qf(x)f(y))).$$

3. *The class of finite directed graphs with a Hamilton cycle.* A *cycle* is a finite structure $\langle A; R \rangle$, where A is a set of n distinct elements $a_1, \cdots a_n$ for some n, and $R = \{\langle a_i, a_{i+1} \rangle : 1 \leqslant i < n\} \cup \{\langle a_n, a_1 \rangle\}$. A *Hamilton cycle* of $\mathfrak{A} = \langle A; G \rangle$ is a cycle $\langle A; H \rangle$, where $H \subseteq G$. This class is a generalized spectrum, via the existential second-order sentence $\exists < \sigma$, where $<$ is a binary predicate symbol, and where σ is the following first-order sentence (which we translate into English for ease in readability):

> "$<$ is a linear order" $\wedge$ "if y is the immediate successor of x in the linear order, then Qxy" $\wedge$ "if x is the minimum element of the linear order and y the maximum, then Qyx."

Our final example is a class of finite structures with a binary function $\circ$.

4. *The class of nonsimple finite groups.* This class is a generalized spectrum, via

$$\exists N (\text{"the structure is a group"} \wedge \text{"}N \text{ is a nontrivial normal subgroup"}).$$

We can ask the generalized Scholz question, as to how to characterize generalized spectra, and the generalized Asser question, as to whether the complement of every generalized spectrum is a generalized spectrum. Of the examples given, it is easy to see that the non-2-colorable finite directed graphs form a generalized spectrum. It is an open question as to whether the complement of any of the others is a generalized spectrum.

It turns out to be possible to characterize spectra and generalized spectra precisely, in terms of time-bounded nondeterministic Turing machines. The concept of a Turing machine is due, of course, to Turing [24]. The concepts of nondeterministic and multi-tape machines are due to Rabin and Scott [17]. The classification by time complexity is due to Hartmanis and Stearns [12], and by tape complexity, to Hartmanis, Lewis and Stearns [11].

In §§2 and 3, we give definitions and background material. Nothing there is new.

In §4, we show the essential equivalence of generalized spectra and nondeterministic polynomial-time recognizable sets. This supplements the known equivalence of spectra and nondeterministic exponential time recognizable sets of positive integers, which is probably due to James Bennett (unpublished); it was also shown by Jones and Selman [15].

In §5, we show, by analyzing our proof of the automata-theoretic characterization of spectra, that many (all?) spectra are the spectrum of a sentence which has at most one model of each finite cardinality.

In §6, we make use of the automata-theoretic characterization of spectra to show that if spectra are not closed under complement, then a class of candidates for counterexamples suggested by Robert Solovay is sufficient.

In §7, we consider Cook's [7] and Karp's [16] notions of polynomial-completeness and reducibility. We generalize to exponential-completeness, and we directly produce (without making use of Cook's or Karp's results) a polynomial-complete set and an exponential-complete set. This was also done by Book [4]; his sets are similar to ours. We show that completeness implies a certain complement-completeness; using this fact, along with our automata-theoretic characterization of generalized spectra, we show that results in Karp's paper [16] (developed by Karp, Tarjan, and Lawler) give us specific examples of generalized spectra whose complements are generalized spectra iff the complement of every generalized spectrum is a generalized spectrum. In particular, we show that the class of finite directed graphs with a Hamilton cycle is such a "complete" generalized spectrum. Also, we find a complete generalized spectrum defined using only one "extra" (existentialized) unary predicate symbol: This is a best possible result. By making use of automata theory and a result about spectra in the author's doctoral dissertation [9], it is shown that there is a complete spectrum defined using only one binary predicate symbol: This is a best possible result.

In §8, we make use of the polynomial-complete set which we constructed in the previous section to show that if the classes of sets which are polynomial-time recognizable by deterministic and nondeterministic Turing machines are the same, then the following apparently much stronger condition holds: There is a constant k such that essentially any set that can be recognized nondeterministically in time T can be recognized deterministically in time T^k. We then generalize this result in various ways. We conclude §8 by an analogy with Post's problem.

In §9, we make use of a tape-complexity argument similar to one used by Bennett [2] to show that there is a spectrum S such that $\{n: 2^n \in S\}$ is not a spectrum. By making use of a result in [9], we then show that there is such a spectrum S defined using only one binary predicate symbol. We also show that our techniques give a new proof of a theorem of Book [3], that the two classes of sets recognizable nondeterministically in polynomial time or in exponential time respectively are different.

In §10, we exhibit an example of a polynomial-complete set which is recognized by a nondeterministic two-tape Turing machine in real time. The existence of such a set follows immediately from theorems of Hunt [14], and of Book and Greibach [5].

2. Notions from logic. Denote the set of *positive integers* $\{1, 2, 3, \cdots\}$ by Z^+, and the set $\{0, \cdots, n-1\}$ by n. By the *natural numbers* we mean the set $Z^+ \cup \{0\}$. If A is a set, then card A is the cardinality of the set. Denote the set of k-tuples $\langle a_1, \cdots, a_k\rangle$ of members of A by A^k.

A *finite similarity type* is a finite set of predicate symbols and function symbols. Each predicate symbol (function symbol) has a positive integer (natural number), the *degree*, associated with it. If a symbol has degree k, then call the symbol k-ary. We will often call 1-ary symbols *unary*, and 2-ary symbols *binary*. A constant symbol is a 0-ary function symbol. We will denote finite similarity types by the letters S and T.

Assume that S contains the n distinct symbols $Q_1, \cdots, Q_n$, written in some fixed order. Then a *finite S-structure* $\mathfrak{A}$ is an $(n+1)$-tuple $\langle A; R_1, \cdots, R_n\rangle$ (where we write a semicolon after the first member), such that we have the following:

1. A is a nonempty finite set, called the *universe* (of $\mathfrak{A}$), and denoted $|\mathfrak{A}|$.
2. If Q_i is a k-ary predicate symbol, then R_i is a subset of A^k.
3. If Q_i is a k-ary function symbol, and $k > 0$, then R_i is a function from A^k into A.
4. If Q_i is a constant symbol, then $R_i \in A$.

In each case, write $R_i = Q_i^{\mathfrak{A}}$. We will sometimes make use of a *graph predicate symbol* Q; if $Q \in S$, then, for $\mathfrak{A}$ to be a finite S-structure, $Q^{\mathfrak{A}}$ must be a graph (i.e., irreflexive and symmetric), or, equivalently, a set of unordered pairs (of members of $|\mathfrak{A}|$). Denote the cardinality of $|\mathfrak{A}|$ by card $(\mathfrak{A})$. Denote the class of finite S-structures by Fin (S); abbreviate Fin $(\{Q_1, \cdots, Q_n\})$ by Fin $(Q_1, \cdots, Q_n)$.

Assume that S and T are disjoint finite similarity types, that $\mathfrak{A}$ is a finite $S \cup T$-structure, and that $\mathfrak{B}$ is a finite S-structure. Then $\mathfrak{A}$ is an *expansion* of $\mathfrak{B}$ (to $S \cup T$) if $|\mathfrak{A}| = |\mathfrak{B}|$ and $Q^{\mathfrak{A}} = Q^{\mathfrak{B}}$ for each Q in S. We write $\mathfrak{B} = \mathfrak{A} \upharpoonright S$.

The metamathematical language we will be working in is a set of symbols $\sim, \wedge, \forall, =$; an infinite number of individual variables u, v, w, x, y, z along with affixes; the left and right parentheses (,); and predicate and function variables. We do not distinguish between predicate or function *symbols* and predicate or function *variables.* Except in this section, whenever we refer to a variable, we will always mean an individual variable.

A *term* is a member of the smallest set T which contains the 0-ary function variables and the individual variables, and which contains $f(t_1, \cdots, t_k)$ for each k-ary function variable f and each $t_1, \cdots, t_k$ in T.

An *atomic formula* is an expression $t_1 = t_2$ or $Qt_1 \cdots t_k$, where the t_i are terms and Q is a k-ary predicate variable. A *first-order formula* is a member of the smallest set which contains each atomic formula, and which contains $\sim \phi_1$, $(\phi_1 \wedge \phi_2)$, and $\forall x \phi_1$ (for each individual variable x), whenever it contains ϕ_1 and ϕ_2. A *second-order formula* is a member of the smallest set which contains each atomic formula, and which contains $\sim \phi_1$, $(\phi_1 \wedge \phi_2)$, $\forall x \phi_1$ (for each individual variable x) and $\forall Q \phi_1$ (for each predicate or function variable Q) whenever it contains ϕ_1 and ϕ_2.

The formulas $\phi_1 \vee \phi_2$, $\exists x \phi$, $(\exists x \neq y)\phi$, $\exists! x \phi$ (read "there exists exactly one x such that ϕ"), and so on, are defined in the usual way, e.g., $\phi_1 \vee \phi_2$ is $\sim(\sim \phi_1 \wedge \sim \phi_2)$. If $T = \{Q_1, \cdots, Q_n\}$ is a finite similarity type, then $\exists T \phi$ is $\exists Q_1 \cdots \exists Q_n \phi$. If ϕ is a first-order formula, then $\exists T \phi$ is called an *existential second-order formula.*

If $x_1, \cdots, x_m$ are individual variables, then we will sometimes write $\mathbf{x}$ as an abbreviation for the m-tuple $\langle x_1, \cdots, x_m \rangle$, when this will lead to no confusion. We may write $\forall \mathbf{x} \phi$ for $\forall x_1 \cdots \forall x_m \phi$.

The notion of a variable being a *free variable* is understood in the usual way. Let S be a fixed finite similarity type. An *S-formula* is a (first- or second-order) formula all of whose free predicate and function variables are in S. A *sentence* is a formula with no free individual variables. A formula is *quantifier-free* if it contains no quantifiers ($\forall$ or $\exists$).

Assume that $\mathfrak{A}$ is a finite S-structure, and that σ is a first- or second-order S-sentence. Then $\mathfrak{A} \models \sigma$ means that σ is true in $\mathfrak{A}$; we say that $\mathfrak{A}$ *is a model of* σ. For a precise definition of truth, see [22]. We note that the equality symbol = is always given the standard interpretation. We define $\mathrm{Mod}_\omega \sigma$ to be the class of all finite S-structures which are models of σ.

Assume that S and T are disjoint finite similarity types, and that $A \subseteq \mathrm{Fin}(S)$. Then A is an *S-spectrum*, or an *(S, T)-spectrum*, if there is a first-order $S \cup T$-sentence σ such that $A = \mathrm{Mod}_\omega \exists T\sigma$. This is simply Tarski's [23] notion of PC, in the special case where we restrict to the class of finite structures. A *generalized spectrum* is an S-spectrum for some S. A *monadic generalized spectrum* is an (S, T)-spectrum where T is a set of unary predicate symbols. A *spectrum* is an S-spectrum for S empty; if A is a spectrum, then we identify $\{n: \langle n \rangle \in A\} \subseteq Z^+$ with A. In this case, if $A = \{n: \langle n \rangle \models \exists T\sigma\}$, then we call A the *spectrum of* σ.

3. Notions from automata theory. When A is a finite set of symbols, then A^* is the set of *strings* or *words*, that is, the finite concatenations $a_1 \frown a_2 \frown \cdots \frown a_n$ of members of A. The *length* of $a = a_1 \frown \cdots \frown a_n$ is n (written $\mathrm{len}(a) = n$). If $k \in Z^+$, then $\mathrm{len}(k)$ is the length of the binary representation of k; this corresponds to a convention that we will always represent positive integers in binary notation. If a set $S \subseteq A^*$ for some finite set A, then S is a *language.*

An *m-tape nondeterministic Turing machine M* is an 8-tuple $\langle K, \Gamma, B, \Sigma, \delta, q_0, q_A, q_R \rangle$, where K is a finite set (the *states* of M); Γ is a finite set (the *tape symbols* of M); B is a member of S (the *blank*); Σ is a subset of $(\Gamma - \{B\})$ (the *input symbols* of M); q_0, q_A, and q_R are members of K (the *initial state, accepting final state,* and *rejecting final state* of M, respectively); and δ is a mapping from $(K - \{q_A, q_R\}) \times \Gamma^m$ to the set of nonempty subsets of $K \times (\Gamma - \{B\})^m \times \{L, R\}^m$ (the table of *transitions, moves,* or *steps* of M).

If the range of δ consists of singletons sets, that is, sets with exactly one member, then M is an *m-tape deterministic Turing machine.*

We may sometimes call M simply a *machine.*

An *instantaneous description* of M is a $(2m + 1)$-tuple $I = \langle q; \alpha^1, \cdots, \alpha^m; i_1, \cdots, i_m \rangle$, where $q \in K$, where $\alpha^j \in (\Gamma - \{B\})^*$, and where $1 \leqslant i_j \leqslant \mathrm{len}(\alpha^j) + 1$, for $1 \leqslant j \leqslant m$. We say that M is in state q, that α^j is the nonblank portion of the jth tape, and that the jth tape head is scanning $(\alpha^j)_{i_j}$, the i_jth symbol of the word α^j (or that M is scanning $(\alpha^j)_{i_j}$ on the jth tape); we also say that the jth tape head is scanning the i_jth tape square.

Let $I' = \langle q'; \alpha^{1'}, \cdots, \alpha^{m'}; i'_1, \cdots, i'_m \rangle$ be another instantaneous description of M. We say that $I \to_M I'$ if $q \neq q_A$, $q \neq q_R$, and if there is $s = \langle p; a_1, \cdots, a_m; T_1, \cdots, T_m \rangle$ in $\delta(q; (\alpha^1)_{i_1}, \cdots, (\alpha^m)_{i_m})$ such that $p = q'$, and, for each j, with $1 \leqslant j \leqslant m$:

1. $(\alpha^{j'})_{i_j} = a_j$.
2. $(\alpha^{j'})_k = (\alpha^j)_k$ for $1 \leqslant k \leqslant \mathrm{len}(\alpha^j)$, if $k \neq i_j$.
3. $\mathrm{len}(\alpha^{j'}) = \mathrm{len}(\alpha^j)$ unless $i_j = \mathrm{len}(\alpha^j) + 1$; in that case, $\mathrm{len}(\alpha^{j'}) = \mathrm{len}(\alpha^j) + 1$.

4. If $T_j = L$, then $i_j \neq 1$.
5. If $T_j = R$, then $i_j' = i_j + 1$; if $T_j = L$, then $i_j' = i_j - 1$.

We say that M prints a_j on the jth tape. Note that M cannot print a blank (that is, $a_j \neq B$); so, we say that α^j is that portion of the jth tape which has been visited, or scanned. If $T_j = R\,(L)$, then we say that the jth tape head moves to the right (left). Assumption 4 corresponds to the intuitive notion of each tape being one-way infinite to the right; thus, if M "orders a tape head to go off the left end of its tape," then M halts. It is important to observe that it is possible to have $I \rightarrow_M I_1$ and $I \rightarrow_M I_2$ with $I_1 \neq I_2$; hence the name "nondeterministic."

We say $I \overset{*}{\rightarrow}_M J$ if there is a finite sequence $I_1, \cdots, I_n$ such that $I_1 = I$, $I_n = J$, and $I_i \rightarrow_M I_{i+1}$ for $1 \leqslant i < n$. Denote the empty word in Σ^* by Λ. If $w \in \Sigma^*$, then let $\bar{w} = \langle q_0; w, \Lambda, \cdots, \Lambda; 1, \cdots, 1\rangle$ (w is the *input*). Call an instantaneous description $\langle q; \alpha^1, \cdots, \alpha^m; i_1, \cdots, i_m\rangle$ *accepting* (*rejecting*) if $q = q_A$ ($q = q_R$). We say that M *accepts* w in Σ^* if $\bar{w} \overset{*}{\rightarrow}_M I$ for some accepting I. Denote by A_M the set of all words accepted by M. We say that M *recognizes* A_M.

If $\bar{w} \overset{*}{\rightarrow}_M I$ for some accepting (rejecting) I, then we say that M, with w as input, eventually enters the accepting (rejecting) final state, and halts.

Intuitively speaking, there are three ways that a word w in Σ^* may be not accepted by M: M, with w as input, can eventually enter the rejecting final state q_R; or M can order a tape head to go off the left end of its tape; or M can never halt.

Assume that M is a multi-tape nondeterministic Turing machine, $w \in A_M$, and t is a positive integer. We say that M *accepts* w *within* t *steps* if, for some $n \leqslant t$,

(2) there are instantaneous descriptions $I_1, \cdots, I_{n+1}$ such that $I_1 = \bar{w}$, I_{n+1} is accepting, and $I_k \rightarrow_M I_{k+1}$ for $1 \leqslant k \leqslant n$.

Let s be a positive integer. Then M *accepts* w *within space* s if for some positive integer n, (2) holds and, for each I_k, $1 \leqslant k \leqslant n+1$, if $I_k = \langle q; \alpha^1, \cdots, \alpha^m; i_1, \cdots, i_m\rangle$, then $i_p \leqslant s$ for $1 \leqslant p \leqslant m$.

Let $T: N \rightarrow N$ and $S: N \rightarrow N$ be functions. We say that M *operates in time* T (*tape* S), or M *recognizes* A_M *in time* T (*tape* S) if, for each natural number l and each word w in A_m of length l, the machine M accepts w within $T(l)$ steps (space $S(l)$). We say that A *is recognizable* (*non*)*deterministically in time* T, or *tape* S, if there is a multi-tape (non)deterministic Turing machine M that operates in time T, or tape S, such that $A = A_M$.

We will now define some well-known, important classes. Let P (NP) be the class of sets A for which there is a positive integer k such that A is recognizable (non)deterministically in time $l \mapsto l^k$. These are the *(non)deterministic polynomial-time recognizable sets.*

Let P_1 (NP_1) be the class of sets A for which there is a positive integer k such that A is recognizable (non)deterministically in time $l \mapsto 2^{kl}$. These are the *(non)deterministic exponential-time recognizable sets.* If the positive integer n has length l in binary notation, then $2^{l-1} \leqslant n < 2^l$. Therefore, a set A of positive integers is in P_1 (NP_1) iff there is a multi-tape (non)deterministic Turing machine M, and a positive integer k such that $A = A_M$ and M accepts each n in A within n^k steps. So in some sense, P_1 and NP_1 are also classes of polynomial time recognizable sets.

We say that a set A is recognizable in *real time* if A is recognizable in time $l \mapsto l + 1$. We use $l + 1$ instead of l, so that the machine can tell when it reaches the end of the input word.

We have defined Turing machines which recognize sets rather than compute functions. It is clear how to modify our definitions to get the usual notion of a function f computable by a deterministic one-tape Turing machine M; it is also clear what we mean by M *computes the value of* f *at* w *within* t *steps.* If $f: A \longrightarrow B$, where A and B are languages, and if $T: N \longrightarrow N$, then we say that M *computes* f *in time* T if, for each natural number l and each word w in A of length l, the machine M computes the value of f at w within $T(l)$ steps. We define Π to be the class of functions which are computable by a one-tape deterministic Turing machine in polynomial time. Functions are generally considered easy to compute if they are in Π; Cobham [6] was the first to single out this class. We define Π_1 to be the class of functions f which are computable by a one-tape deterministic Turing machine in exponential time, and for which there is a constant c such that $\text{len}(f(w)) \leqslant c \cdot \text{len}(w)$ for each w in the domain of f.

We now state without proof two theorems, which were essentially proved in [12]. The proofs can also be found in [13, pp. 139–140 and 143].

THEOREM 1. *If* A *is recognized by a one-tape (non)deterministic Turing machine in time* T, *if* $\liminf_{l\to\infty} T(l)/l^2 = \infty$, *and if* $c > 0$ *is arbitrary, then* A *is recognized by a one-tape (non)deterministic Turing machine in time* $l \mapsto \max(l + 1, cT)$.

THEOREM 2. *If* A *is recognized by an m-tape (non)deterministic Turing machine in time* T, *and if* $\liminf_{l\to\infty} T(l)/l = \infty$, *then* A *is recognized by a one-tape (non)deterministic Turing machine in time* T^2.

It follows from Theorem 2 that the concepts of polynomial and exponential time are invariant, whether we consider one-tape or multi-tape machines.

A function T is *countable* if there is a positive integer c and a two-tape deterministic Turing machine that operates in time cT which, for each natural number l and each word of length l as input (on the first tape), halts (by entering a final state) with a string of at least $T(l)$ tallies on its second tape (a tally is a one). This is slightly broader than the usual definition, but more convenient for us to use. We will make use of the fact that $l \longmapsto l^k$ is countable for each positive integer k; for, l^k can be calculated in a polynomial of $\operatorname{len}(l)$ time, which is less than l for sufficiently large l.

A *linear-bounded automaton* is a one-tape deterministic Turing machine that operates in tape $l \longmapsto l + 1$. We denote by $\mathcal{E}_*^2$ those subsets of Z^+ whose characteristic functions are in the Grzegorczyk class $\mathcal{E}^2$ [10]. The class $\mathcal{E}^2$ is the smallest class which contains the successor and multiplication functions, and is closed under explicit transformation, composition, and limited recursion. We are interested in the class $\mathcal{E}_*^2$ precisely because of the following theorem.

THEOREM 3 (RITCHIE [18]). *A set of positive integers is recognizable by a linear-bounded automaton iff it is in* $\mathcal{E}_*^2$.

We will make use of the following well-known simple theorem, which we state without proof.

THEOREM 4. *The classes* $\mathcal{E}_*^2$, $\mathcal{P}$, *and* $\mathcal{P}_1$ *are closed under complement.*

A function $S: N \longrightarrow N$ is said to be *constructible* if there is a one-tape deterministic Turing machine which operates in tape S, but not in tape S', if $S'(l) < S(l)$ for some l. We conclude this section by stating a theorem which is essentially proved in [11]. The proof can also be found in [13, pp. 150–151].

THEOREM 5. *Assume that* S *is a constructible tape function with* $S(l) \geqslant \log_2 l$ *for each natural number* l. *Then there is a set which is recognizable by a one-tape deterministic Turing machine in tape* S, *but which is not recognizable in tape* S' *for any function* S' *with* $\liminf_{l \to \infty} S'(l)/S(l) = 0$.

4. Generalized spectra and automata. In this section, we will prove the theorem (Theorem 6) which interrelates spectra and generalized spectra with automata.

Let S be a fixed finite similarity type which (for convenience) contains only predicate symbols, and let $P_1, \cdots, P_r$ be the predicate symbols in S in some fixed order. Let $\Sigma = \{0, 1, \#\}$.

Assume that $\mathfrak{A} = \langle\{1, \cdots, n\}; S_1, \cdots, S_r\rangle$ is a finite S-structure, and that P_i (and hence S_i) is m_i-ary $(1 \leq i \leq r)$. For each i, define b_i to be the word in $\{0, 1\}^*$ of length n^{m_i} such that if $\langle c_1, \cdots, c_{m_i}\rangle$ is the kth member of $\{1, \cdots, n\}^{m_i}$ in lexicographical order, then the kth digit of b_i is 1 if $S_i c_1 \cdots c_{m_i}$ and 0 otherwise $(1 \leq k \leq n^{m_i})$. Let $e(\mathfrak{A})$, the *encoding* of $\mathfrak{A}$, be the word $a \# b_1 \# b_2 \# \cdots \# b_r$ in Σ^*, where a is the binary representation of n. If A is a class of finite $\mathcal{S}$-structures, then let $E(\mathsf{A}) = \{e(\mathfrak{A}): (\exists n \in Z^+)(|\mathfrak{A}| = \{1, \cdots, n\}$ and $\mathfrak{A} \in \mathsf{A})\}$.

THEOREM 6. *Assume that* $\mathsf{A} \subseteq \mathrm{Fin}(\mathcal{S})$, *and that* A *is closed under isomorphism.*

1. *If* $\mathcal{S} \neq \emptyset$, *then* A *is an* $\mathcal{S}$-*spectrum iff* $E(\mathsf{A}) \in \mathcal{NP}$.
2. *If* $\mathcal{S} = \emptyset$, *then* A *is a spectrum iff* $E(\mathsf{A}) \in \mathcal{NP}_1$.

Note. We can combine these last two statements by saying that A is an $\mathcal{S}$-spectrum $(\mathcal{S} = \emptyset$ or $\mathcal{S} \neq \emptyset)$ iff there is a positive integer k and there is a nondeterministic multi-tape Turing machine which recognizes $E(\mathsf{A})$, and which accepts each $e(\mathfrak{A})$ in $E(\mathsf{A})$ within n^k steps, where $|\mathfrak{A}| = \{1, \cdots, n\}$.

PROOF. Assume that A is an $\mathcal{S}$-spectrum (possibly $\mathcal{S} = \emptyset$.) Then, for some positive integers t and k, some set $\mathcal{T}$ of t new k-ary predicate symbols, and some first-order $\mathcal{S} \cup \mathcal{T}$-sentence $\sigma = Q_1 x_1 \cdots Q_m x_m \phi$, where ϕ is quantifier-free and each Q_i is $\forall$ or $\exists$, we have $\mathsf{A} = \mathrm{Mod}_\omega \exists \mathcal{T}\sigma$. This is because of well-known techniques of simulating $(k-1)$-ary functions and $(k-1)$-ary relations by k-ary relations, and because each first-order sentence is equivalent to a sentence with all quantifiers out front (so-called "prenex normal form").

We will informally describe a $(t + m + 2)$-tape nondeterministic Turing machine M which recognizes $E(\mathsf{A})$. The first tape is the input tape. The machine M first tests to see if the input is of form $a \# b_1 \# b_2 \# \cdots \# b_r$, with a in $\{0, 1\}^*$ and starting with a 1, with r the number of (predicate) symbols in $\mathcal{S}$, and with each b_i in $\{0, 1\}^*$ and of the proper length; to test for proper length, M uses its last tape as a "counter." If the input is not of the proper form, then M rejects. If the input is of the proper form, then say the input is $e(\mathfrak{A})$, and $|\mathfrak{A}| = \{1, \cdots, n\}$. On each of the 2nd through $(t + 1)$st tapes, M then nondeterministically prints a string of n^k 0's and 1's, by using the last tape as a counter; these correspond to "guesses" for the interpretations of the predicate symbols in $\mathcal{T}$. Let $\mathfrak{A}'$ be the obvious expansion of $\mathfrak{A}$ to $\mathcal{S} \cup \mathcal{T}$.

Next, on the $(t + i + 1)$st tape, M systematically prints each possibility a_i for x_i $(1 \leq i \leq m)$, where a_i runs between 1 and n. There are n^m possibilities for the m-tuple $\langle a_1, \cdots, a_m\rangle$. For each given such possibility, M can easily test to see if $\phi(a_1, \cdots, a_m)$ holds in $\mathfrak{A}'$, where it again makes use of the last tape as a work tape. It is easy to see how to arrange the logic to test whether $\mathfrak{A}' \models \sigma$.

So M recognizes $E(A)$, and there is a nonnegative polynomial p such that M accepts each $e(\mathfrak{A})$ in $E(A)$ nondeterministically within $p(n)$ steps, where $|\mathfrak{A}| = \{1, \cdots, n\}$. Let l be the length of the input $e(\mathfrak{A})$. If $S = \emptyset$, then n is approximately 2^l. If $S \neq \emptyset$, then l is approximately tn^k (in each case, "approximately" means up to a fixed constant factor). So if $S = \emptyset$, then $E(A)$ can be recognized nondeterministically in time $l \mapsto p(2^l)$, and hence $E(A) \in NP_1$. If $S \neq \emptyset$, then $E(A)$ can certainly be recognized nondeterministically in time $l \mapsto p(l)$, and hence $E(A) \in NP$.

Conversely, assume that $E(A)$ is in NP or NP_1, depending on whether $S \neq \emptyset$ or $S = \emptyset$. Assume that $S = \{P_1, \cdots, P_r\}$, where P_i is m_i-ary, for $1 \leq i \leq r$. It will be convenient to define a slightly modified $(r + 1)$-tape nondeterministic Turing machine M, by changing the definition of an input. If x is an $(r + 1)$-tuple $\langle a_1, \cdots, a_{r+1}\rangle$, then let $\bar{x} = \langle q_0; a_1, \cdots, a_{r+1}; 1, \cdots, 1\rangle$; we say that M accepts the $(r + 1)$-tuple x if $\bar{x} \xrightarrow[M]{*} I$ for some accepting instantaneous description I.

It is clear that there is a positive integer k and a modified $(r + 1)$-tape nondeterministic Turing machine M which accepts precisely those $(r + 1)$-tuples $\langle a', b_1, \cdots, b_r\rangle$ such that $a \# b_1 \# b_2 \# \cdots \# b_r$ is in $E(A)$, where a' is the string a written backwards, and such that M accepts such an input within n^k steps, where n is the number represented by a in binary notation. We can assume that $k \geq \max\{m_i : 1 \leq i \leq r\}$. It is clear that if M accepts $\langle a', b_1, \cdots, b_r\rangle$ within n^k steps, then it accepts $\langle a', b_1, \cdots, b_r\rangle$ within space n^k; we will make use of this fact.

Introduce the set T of the following new symbols. The symbol $<$ is a binary predicate symbol, which represents a linear order; c_1 and c_2 are constant symbols, which represent respectively the minimal and maximal members of the linear order; 0, 1, and B are constant symbols, which represent respectively the zero, one and blank tape symbols; q_0, q_A, and q_R are constant symbols, which represent respectively the initial state, accepting final state, and rejecting final state; S is a unary function symbol, which represents successor in the linear order $<$; S_1 is a $2k$-ary predicate symbol, where

$$S_1(x_1, \cdots, x_k; y_1, \cdots, y_k)$$

means that $\mathbf{y}$ is the successor of $\mathbf{x}$ in the lexicographical ordering; q is a k-ary function symbol, where $q(t_1, \cdots, t_k)$ is the state that the machine is in at time $\mathbf{t}$; v_i is a $2k$-ary function symbol, for $1 \leq i \leq r + 1$, where $v_i(t_1, \cdots, t_k; x_1, \cdots, x_k)$ is the tape symbol printed on square $\mathbf{x}$ of the ith tape at time $\mathbf{t}$; H_i is a $2k$-ary predicate symbol, for $1 \leq i \leq r + 1$, where $H_i(\mathbf{t}; \mathbf{x})$ means that, at time $\mathbf{t}$, the ith tape head is scanning square $\mathbf{x}$ on the

ith tape; and G is a binary function symbol, where $G(x, i)$ is the ith digit from the right in the binary representation of x, if we think of the binary representation of the positive integers less than or equal to n (the cardinality of the universe) as being given by a word of length n, which starts out with a series of blanks, followed by the usual binary representation.

We think of the k-tuple $\langle c_1, \cdots, c_1 \rangle$ as representing the first time unit (and the first tape square on each tape); if $S_1(\mathbf{x}; \mathbf{y})$, then $\mathbf{y}$ is the next time unit (next tape square) after $\mathbf{x}$. Thus, the k-tuple $\langle c_2, \cdots, c_2 \rangle$ represents the n^kth time unit (n^kth tape square).

Assume that Γ contains g tape symbols. We represent these by c_1, $S(c_1)$, $S(S(c_1)), \cdots, S^{(g-1)}(c_1)$, where c_1 represents the zero, $S(c_1)$ represents the one, and $S^{(2)}(c_1)$ represents the blank. For ease in readability, we have introduced the symbols 0, 1, and B, which will denote the same elements (in a model) as c_1, $S(c_1)$, and $S^{(2)}(c_1)$ respectively. Assume that K contains p states. We represent these by $c_1, \cdots, S^{(p-1)}(c_1)$ where, for ease in readability, we have q_0, q_A, and q_R denoting the same elements as c_1, $S(c_1)$, and $S^{(2)}(c_1)$ respectively.

Let σ_1 be the conjunction of the following sentences:

$$\begin{aligned} 0 &= c_1, & q_0 &= c_1, \\ 1 &= S(c_1), & q_A &= S(c_1), \\ B &= S^{(2)}(c_1), & q_R &= S^{(2)}(c_1). \end{aligned}$$

Let σ_2 be the sentence "$<$ is a linear order, c_1 is minimal, c_2 is maximal, and S is successor, except $S(c_2) = c_1$."

Let σ_3 be the sentence which says that $S_1(\mathbf{x}; \mathbf{y})$ holds iff $\mathbf{y}$ is the successor of $\mathbf{x}$ in lexicographical order, except that $S_1(c_2, \cdots, c_2; c_1, \cdots, c_1)$ holds. Thus, σ_3 is the conjunction of the following $k + 2$ sentences:

$$\forall x_1 \cdots \forall x_k \exists! y_1 \cdots \exists! y_k S_1(x_1, \cdots, x_k; y_1, \cdots, y_k),$$

$$\forall x_1 \cdots \forall x_k (x_k \neq c_2 \rightarrow S_1(x_1, \cdots, x_k; x_1, \cdots, x_{k-1}, Sx_k)),$$

$$\forall x_1 \cdots \forall x_k ((x_k = c_2 \wedge x_{k-1} \neq c_2)$$
$$\rightarrow S_1(x_1, \cdots, x_k; x_1, \cdots, x_{k-2}, Sx_{k-1}, c_1)),$$

$$\cdots$$

$$S_1(c_2, \cdots, c_2; c_1, \cdots, c_1).$$

The conjunction σ_4 of the following sentences defines G to be what we said we wanted:

$$G(c_1, c_1) = 1,$$

$$(\forall x \neq c_1)(G(c_1, x) = B),$$

$$(\forall x \neq c_2)\forall y(((\exists z < y)(G(x, z) = 0 \vee G(x, z) = B)) \rightarrow (G(Sx, y) = G(x, y))),$$

$$(\forall x \neq c_2)\forall y(((\exists z < y)(G(x, z) = 1 \wedge G(x, y) = 0)) \rightarrow (G(Sx, y) = 1)),$$

$$(\forall x \neq c_2)\forall y(((\forall z < y)(G(x, z) = 1 \wedge G(x, y) = 1)) \rightarrow (G(Sx, y) = 0)),$$

$$(\forall x \neq c_2)\forall y(((\forall z < y)(G(x, z) = 1 \wedge G(x, y) = B)) \rightarrow (G(Sx, y) = 1)).$$

The conjunction σ_5 of the following sentences gives self-explanatory information about q and the H_i:

$$q(c_1, \cdots, c_1) = q_0,$$

$$q(c_2, \cdots, c_2) = q_A,$$

$$\bigwedge_{i=1}^{r+1} \forall t_1 \cdots \forall t_k \exists! x_1 \cdots \exists! x_k H_i(\mathbf{t}; \mathbf{x}),$$

$$\bigwedge_{i=1}^{r+1} H_i(c_1, \cdots, c_1; c_1, \cdots, c_1).$$

The conjunction σ_6 of the next two sentences initializes the first tape so that it starts with the binary representation of n (the cardinality of the universe) running backwards, followed by blanks:

$$\forall x(v_1(c_1, \cdots, c_1; c_1, \cdots, c_1, x) = G(c_2, x)),$$

$$\forall x_1 \cdots \forall x_k(\sim (x_1 = c_1 \wedge \cdots \wedge x_{k-1} = c_1) \rightarrow (v_1(c_1, \cdots, c_1; x_1, \cdots, x_k) = B)).$$

The conjunction σ_7 of the following sentences initializes the 2nd through $(r + 1)$st tapes such that the $(i + 1)$st tape starts out with a string of 0's and 1's which represents P_i, followed by blanks $(1 \leq i \leq r)$:

$$\bigwedge_{i=1}^{r} \forall x_1 \cdots \forall x_{m_i}(P_i x_1 \cdots x_{m_i} \rightarrow (v_{i+1}(c_1, \cdots, c_1; c_1, \cdots, c_1, x_1, \cdots, x_{m_i}) = 1)),$$

$$\bigwedge_{i=1}^{r} \forall x_1 \cdots \forall x_{m_i}(\sim P_i x_1 \cdots x_{m_i} \rightarrow (v_{i+1}(c_1, \cdots, c_1; c_1, \cdots, c_1, x_1, \cdots, x_{m_i}) = 0)),$$

$$\bigwedge_{i=1}^{r} \forall x_1 \cdots \forall x_k(\sim (x_1 = c_1 \wedge \cdots \wedge x_{k-m_i} = c_1) \rightarrow (v_{i+1}(c_1, \cdots, c_1; x_1, \cdots, x_k) = B)).$$

The sentence σ_8 says that after the machine enters a final state, nothing ever changes. Here $\mathbf{u}$ represents the next time unit after $\mathbf{t}$:

$$\forall \mathbf{t} \forall \mathbf{u} \Bigg(\sim (t_1 = c_2 \wedge \cdots \wedge t_k = c_2) \wedge S_1(\mathbf{t}; \mathbf{u}) \wedge (q(\mathbf{t}) = q_A \vee q(\mathbf{t}) = q_R) \rightarrow \Bigg((q(\mathbf{u}) = q(\mathbf{t})) \wedge \forall \mathbf{x} \Bigg(\bigwedge_{i=1}^{r+1} (v_i(\mathbf{u}; \mathbf{x}) = v_i(\mathbf{t}; \mathbf{x})) \Bigg) \wedge \forall \mathbf{x} \bigwedge_{i=1}^{r+1} (H_i(\mathbf{u}; \mathbf{x}) \leftrightarrow H_i(\mathbf{t}; \mathbf{x})) \Bigg) \Bigg).$$

The sentence σ_9 is a conjunction of sentences which describe the table of transitions of M, entry by entry. Assume that $\delta(b; e_1, \cdots, e_{r+1}) = \{s_1, \cdots, s_w\}$, that we are representing the state b by $S^{(d)}(c_1)$, and that we are representing the tape symbol e_i by $S^{(f_i)}(c_1)$, for $1 \leqslant i \leqslant r$. Then one conjunct of σ_9 is the following sentence:

$$\forall \mathbf{t} \forall \mathbf{u} \forall x_1^1 \cdots \forall x_k^1 \cdots \forall x_1^{r+1} \cdots \forall x_k^{r+1} \Bigg(\sim (t_1 = c_2 \wedge \cdots \wedge t_k = c_2) \wedge S_1(\mathbf{t}; \mathbf{u}) \wedge \bigwedge_{i=1}^{r+1} H_i(\mathbf{t}; x_1^i, \cdots, x_k^i) \wedge (q(\mathbf{t}) = S^{(d)}(c_1)) \wedge \bigwedge_{i=1}^{r+1} (v_i(\mathbf{t}; x_1^i, \cdots, x_k^i) = S^{(f_i)}(c_1)) \rightarrow \bigvee_{i=1}^{w} \phi_i \Bigg),$$

where ϕ_i tells the transition which is possible, according to s_i, for $1 \leqslant i \leqslant w$.

Specifically, assume that s_i is $\langle a; b_1, \cdots, b_{r+1}; T_1, \cdots, T_{r+1}\rangle$, where we are representing the state a by $S^{(m)}(c_1)$, where we are representing the symbol b_j by $S^{(d_j)}(c_1)$, for $1 \leqslant j \leqslant r+1$, and where each T_j is either R or L. Let $I = \{j: T_j = R\}$, and $J = \{j: T_j = L\}$. Then ϕ_i is the conjunction of the following formulas, where in the last conjunction we include the restriction that no tape head go off the left end of its tape:

$$q(\mathbf{u}) = S^{(m)}(c_1),$$

$$\bigwedge_{j=1}^{r+1} \forall \mathbf{z}(\sim (z_1 = x_1^j \wedge \cdots \wedge z_k = x_k^j) \longrightarrow (v_j(\mathbf{u}; \mathbf{z}) = v_j(\mathbf{t}; \mathbf{z}))),$$

$$\bigwedge_{j=1}^{r+1} v_j(\mathbf{u}; \mathbf{x}^j) = S^{(d_j)}(c_1), \qquad \bigwedge_{j \in I} \forall \mathbf{y}^j (S_1(\mathbf{x}^j; \mathbf{y}^j) \longrightarrow H_j(\mathbf{u}; \mathbf{y}^j)),$$

$$\bigwedge_{j \in J} (\sim (x_1^j = c_1 \wedge \cdots \wedge x_k^j = c_1) \wedge \forall \mathbf{y}^j (S_1(\mathbf{y}^j; \mathbf{x}^j)) \longrightarrow H_j(\mathbf{u}; \mathbf{y}^j)).$$

If $n \geqslant \max(\text{card } \Gamma, \text{card } K)$, then an S-structure $\mathfrak{A}$ with card $(\mathfrak{A}) = n$ is in A iff $\mathfrak{A} \models \exists T(\bigwedge_{i=1}^{9} \sigma_i)$. It is well known that each "finite modification" of an S-spectrum is an S-spectrum. Therefore, A is an S-spectrum. □

Apparently, James Bennett was the first to prove part 2 of Theorem 6, although he did not publish it. The first published proof (a different proof from ours) is by Jones and Selman [15]. Part 1 is new.

It is fairly easy to prove from Theorem 6 that

(3) the class of (generalized) spectra is closed under complement iff NP_1 (NP) is closed under complement.

This is because there are not only simple ways to encode finite structures into strings of symbols, but also ways to "encode" strings of symbols into finite structures. We will not demonstrate this, because (3) follows easily from our work on complete sets in §7.

We know from Theorem 4 that P_1 (P) is closed under complement. So if $NP_1 = P_1$ ($NP = P$), then NP_1 (NP) is closed under complement, and hence the class of (generalized) spectra is closed under complement. It is a famous open problem in automata theory as to whether $NP = P$; the evidence seems to be strongly against it. We remark that it is well known that $NP = P$ implies that $NP_1 = P_1$, and that if NP is closed under complement, then so is NP_1; these results follow, for example, by an obvious modification of an argument by Savitch [20, p. 186].

From Theorem 6, we see that spectra and generalized spectra are very broad classes. Most sets of positive integers that occur in number theory, such as the primes, the Fibonacci numbers, and the perfect numbers, are easily seen to be members of $\mathcal{P}_1$, and a fortiori of $N\mathcal{P}_1$. It is immediate from Theorem 6(2) that a set of positive integers is in $N\mathcal{P}_1$ iff it is a spectrum.

THEOREM 7 (BENNETT [2]). *Assume that the set A of positive integers is in $\mathcal{E}^2_*$. Then A and $\widetilde{A}$ are spectra.*

PROOF. By Theorem 3, A is recognizable by a linear-bounded automaton. So, by an easy, standard argument of counting the number of possible instantaneous descriptions, it follows that $A \in \mathcal{P}_1$. So $A \in N\mathcal{P}_1$, and hence A is a spectrum. Since $\mathcal{E}^2_*$ is closed under complement (Theorem 4), also $\widetilde{A} \in \mathcal{E}^2_* \subseteq \mathcal{P}_1 \subseteq N\mathcal{P}_1$; hence $\widetilde{A}$ is a spectrum. □

It is an open problem as to whether there is any spectrum not in $\mathcal{E}^2_*$.

Let BIN be the set of all spectra definable using only one graph predicate symbol. Obviously, if $S \in$ BIN, then S is definable using only one binary predicate symbol. The following result is proved in the author's doctoral dissertation [9].

THEOREM 8. *For each spectrum S, there is a positive integer k such that $\{n^k: n \in S\}$ is in BIN.*

We could not hope for it to be true that for each spectrum S, there is a positive integer k such that $\{n^k: n \in S\}$ is definable using only unary predicate symbols. This is because it is well known that by an elimination-of-quantifiers argument, it can be shown that each spectrum definable using only unary predicate symbols is either a finite or a cofinite set of positive integers.

We close this section by using Theorem 8 to show that if certain conjectures about spectra are false, then a counterexample occurs in BIN.

THEOREM 9. *The following two statements are equivalent.*

1. $N\mathcal{P}_1 = \mathcal{P}_1$.
2. $BIN \subseteq \mathcal{P}_1$.

PROOF. $1 \Rightarrow 2$: BIN $\subseteq N\mathcal{P}_1$, by Theorem 6(2).

$2 \Rightarrow 1$: Take S in $N\mathcal{P}_1$; we want to show that $S \in \mathcal{P}_1$. By the usual encoding arguments, we can assume that $S \subseteq Z^+$. By Theorem 8, we can find a positive integer k such that $T = \{n^k: n \in S\}$ is in BIN. Then $n \in S$ iff $n^k \in T$, for each positive integer n. So clearly, if $T \in \mathcal{P}_1$, then $S \in \mathcal{P}_1$. □

THEOREM 10. *The following two statements are equivalent.*

1. *If* $S \subseteq Z^+$, *then* $S \in NP_1$ *iff* $S \in E_2^*$.
2. $BIN \subseteq E_*^2$.

The proof is very similar to the previous proof. □

5. Categoricity. Call a first-order sentence *categorical* if it has at most one model (up to isomorphism) of each finite cardinality.

THEOREM 11. *Assume that the set* S *of positive integers is in* P_1. *Then there is a categorical sentence with spectrum* S.

PROOF. If the machine M in the proof of Theorem 6 is deterministic, then the sentence $\bigwedge_{i=1}^{9} \sigma_i$ defined in that proof is categorical. The "finite modification" which was called for to take care of small values of n is easily dealt with. □

So those naturally-occurring sets of positive integers that we discussed in the previous section are each the spectrum of a categorical sentence.

COROLLARY 12. *If* $NP_1 = P_1$, *then each spectrum is the spectrum of a categorical sentence.*

The proof is immediate. □

In the case of S-spectra, let us call a first-order $S \cup T$-sentence σ (where $S \cap T = \emptyset$) *S-categorical* if, whenever $\mathfrak{A}$ and $\mathfrak{B}$ are finite $S \cup T$-structures which are models of σ, and $\mathfrak{A} \upharpoonright S$ and $\mathfrak{B} \upharpoonright S$ are isomorphic, then so are $\mathfrak{A}$ and $\mathfrak{B}$.

If A is an S-spectrum, then it does not quite follow, as in Theorem 11, that if $E(A) \in P$ then there is T and there is an S-categorical $S \cup T$-sentence σ such that $A = \mathrm{Mod}_\omega \exists T \sigma$. For, there are many different ways to impose the linear ordering $<$. However, if structures had a "built-in" linear ordering, then we could surmount this difficulty. One approach is to consider only finite S-structures $\mathfrak{A}$ such that $|\mathfrak{A}| \subseteq Z^+$. We could let $<$ be a symbol which, like $=$, has an invariant interpretation; namely, if $a, b \in |\mathfrak{A}|$, where $|\mathfrak{A}| \subseteq Z^+$, then $a <^{\mathfrak{A}} b$ iff a is a smaller integer than b. Then the desired result mentioned above follows.

6. Possible Asser counterexamples. In §1, we gave several simple examples of generalized spectra whose complements do not seem to be generalized spectra. These also serve as examples of NP sets which are probably not in P.

It is harder to find candidates for sets which are spectra but whose complements are not spectra, or which are in NP_1 but not in P_1. This is because, as we observed, most naturally-occurring sets of positive integers are in P_1, and hence, of course, so are their complements.

As we shall see in §9, there exists a spectrum S such that $\{n: 2^n \in S\}$ is not a spectrum. This gives us one class of possible counterexamples. In fact, Bennett [2] shows that $\{n: 2^n + 1$ is composite$\}$ is a spectrum, and asks whether $\{n: 2^n + 1$ is prime$\}$ is a spectrum. We will show that Bennett's result follows fairly simply from Theorem 6 (Bennett's proof is different). We will answer Bennett's question (affirmatively) by making use of a very surprising result by Vaughn Pratt (unpublished). We need the following simple theorem.

THEOREM 13. *Assume* $A \subseteq Z^+$. *If* $A \in NP$, *then* $\{n: 2^n + 1 \in A\} \in NP_1$.

PROOF. Assume that M is a nondeterministic Turing machine which recognizes A in polynomial time. We will define a nondeterministic Turing machine M' which recognizes $B = \{n: 2^n + 1 \in A\}$ in exponential time. Given input n, the machine M' prints the string that starts and ends with a 1 and has $(n - 1)$ 0's in between. This is the number $2^n + 1$ in binary notation. Then M' simulates the action of M on input $2^n + 1$. It is easy to see that M' recognizes B nondeterministically in exponential time. □

It is simple to show that $C = \{n: n$ is composite$\}$ is in NP. For, let M be a nondeterministic Turing machine which, given input n, "guesses" a divisor m of n and then tests it; if m divides n, then M accepts n. Clearly M recognizes C nondeterministically in polynomial time. So, from Theorem 13, $\{n: 2^n + 1$ is composite$\}$ is in NP_1, and hence is a spectrum.

Pratt proved that $\{n: n$ is prime$\}$ is in NP. From this very interesting result, it follows immediately from Theorem 13 that $\{n: 2^n + 1$ is prime$\}$ is a spectrum.

For each set S of words, define S' to be $\{\text{len}(n): n \in S\}$. As candidates for sets in NP_1 which are not in P_1, Robert Solovay (personal communication) essentially suggested considering sets S', where $S \in P$. We will now show that in a certain sense, this class is sufficient for a counterexample. The proof gives an application to automata theory of the equivalence in Theorem 6.

THEOREM 14. *The following three statements are equivalent*:

1. $NP_1 = P_1$.
2. *If* $S \in P$, *then* $S' \in P_1$.
3. *If* $S \in NP$, *then* $S' \in P_1$.

PROOF. $3 \Rightarrow 2$: This is immediate, since $P \subseteq NP$.

$1 \Rightarrow 3$: Assume that $S \in NP$. Then $S' \in NP_1$. For, assume that M recognizes S nondeterministically in time $l \mapsto l^k$, for some k. We will construct a machine M' that recognizes S' nondeterministically in exponential time. Given input m, the machine M' first guesses a number n of length m.

Then M' simulates M on the input n. Clearly, M' recognizes S'; M' accepts m in S' in roughly m^k steps. So $S' \in NP_1 = P_1$.

$2 \Rightarrow 1$: Assume that $A \in NP_1$. By the usual encoding arguments, we can assume that $A \subseteq Z^+$ (if, instead, $A \subseteq \Sigma^*$ for the finite set Σ, then we find an encoding $t: \Sigma^* \longrightarrow Z^+$ for which there is a constant c such that $\text{len}(t(w)) = c \cdot \text{len}(w)$ for each w in Σ^*; the details are straightforward). By Theorem 6, we know that A is a spectrum. Assume for simplicity that $A = \{n: \langle n \rangle \models \exists Q\sigma\}$, where Q is a binary predicate symbol. The general case is similar. Let S be the following set:

$$\{m: \exists n (\text{len}(m) = n^2 + 1, \text{ and if the binary representation of } m \text{ is } 1 \frown b, \text{ and if } R \text{ is the binary relation on } n \text{ which is represented by } b \text{ in the obvious way, then } \langle n; R \rangle \models \sigma)\}.$$

We use $n^2 + 1$ instead of n^2 to allow for the possibility of b being a string of all 0's.

Then $S \in P$. For, as we saw in the proof of Theorem 6, there is a positive integer k and a deterministic machine M which can determine whether $\langle n; R \rangle \models \sigma$ within n^k steps for each n (and R); note that n^k is bounded by a fixed polynomial of the length of m.

Since $S \in P$, it follows by hypothesis that $S' \in P_1$. Now $n \in A$ iff $(n^2 + 1) \in S'$, for each positive integer n. So $A \in P_1$. □

In the next section, we will find several specific (generalized) spectra A ($\mathcal{A}$) such that the class of (generalized) spectra is closed under complement iff $\widetilde{A}$ ($\widetilde{\mathcal{A}}$) is a (generalized) spectrum.

7. Complete sets. We will now deal with the notions of reducibility and completeness, which are due to Cook [7] and Karp [16]. We will show that completeness implies a complement-completeness (Theorem 15(2)), and we will use this fact, along with Theorem 6(1) and results in Karp's paper [16], to find particular generalized spectra whose complements are generalized spectra iff the complement of every generalized spectrum is generalized spectrum. In particular, we will show that the class of directed graphs with a Hamilton cycle is such a "complete" generalized spectrum; we will also exhibit a monadic complete generalized spectrum. We will then find a complete spectrum, and will show that the existence of a complete spectrum implies the existence of a complete spectrum in BIN (which we can actually find).

Assume that Σ_1 and Σ_2 are finite sets, and $A \subseteq \Sigma_1^*, B \subseteq \Sigma_2^*$. *$B$ is reducible* (*reducible*$_1$) *to* A, written $B \propto A$ ($B \propto_1 A$), if there is a function f: $\Sigma_2^* \longrightarrow \Sigma_1^*$ in Π (Π_1) such that, for each x in Σ_2^*, $x \in B$ iff $f(x) \in A$.

It is clear that $\propto$ and $\propto_1$ are transitive.

A set A is NP (NP_1)*-complete* if

1. $A \in NP$ (NP_1).
2. $B \propto A$ $(B \propto_1 A)$ for each B in $NP(NP_1)$.

THEOREM 15. *Let A be $NP(NP_1)$-complete. Then*

1. $NP = P$ $(NP_1 = P_1)$ *iff* $A \in P(P_1)$;
2. $NP(NP_1)$ *is closed under complement iff* $\widetilde{A} \in NP(NP_1)$.

PROOF. Assume that $B \subseteq \Sigma^*$, that $B \in NP(NP_1)$ and that $B \propto A$ $(B \propto_1 A)$. Find f in Π (Π_1) such that $x \in B$ iff $f(x) \in A$, for each x in Σ^*. It is straightforward to check that if $A \in P(P_1)$, then $B \in P(P_1)$. Note that $x \in \widetilde{B}$ iff $f(x) \in \widetilde{A}$; hence, if $\widetilde{A} \in NP(NP_1)$, then $\widetilde{B} \in NP(NP_1)$. The other implications are obvious. □

Part 1 of Theorem 15 (in the $NP = P$ case) is due to Karp. Cook was the first to show that there exists an NP-complete set. This set is SAT, the set of encodings of satisfiable propositional formulas in "conjunctive normal form" $\bigwedge_i \bigvee_j A_{ij}$, where each A_{ij} is a propositional letter or its negation.

THEOREM 16 (COOK [7]). *SAT is NP-complete.*

In [16], SAT is shown to be reducible to certain other sets in NP, which are thus NP-complete. We now describe two such sets.

1. HAM is the set of encodings of $\{Q\}$-structures that have a Hamilton cycle, where Q is a binary predicate symbol.

2. HIT is the set of encodings of families of subsets of a set, for which there is a "hitting set." If the input is (the encoding of) a finite family $\{A_1, \cdots, A_n\}$, where each $A_i \subseteq \{s_1, \cdots, s_r\}$, then a hitting set is a set $W \subseteq \{s_1, \cdots, s_r\}$ such that $W \cap A_i$ contains exactly one element for each i.

THEOREM 17 (KARP, TARJAN, AND LAWLER [16]). *HAM and HIT are NP-complete.*

We can now demonstrate two particular generalized spectra whose complements are generalized spectra iff the complement of every generalized spectrum is a generalized spectrum. Let Q be a binary predicate symbol, and U a unary predicate symbol.

THEOREM 18. *Let* $\mathsf{A} = \{\mathfrak{A} \in \mathrm{Fin}(Q)\colon \mathfrak{A}$ *has a Hamilton cycle*$\}$. *Then the class of generalized spectra is closed under complement iff the complement of the $\{Q\}$-spectrum A is a $\{Q\}$-spectrum.*

PROOF. $\Rightarrow$: This is immediate.

$\Leftarrow$: This would follow immediately from Theorems 6(1), 15(2) and 17 except for one technicality. Namely, if $\mathcal{B}$ is an $\mathcal{S}$-spectrum, and if C is the complement of $\mathcal{B}$ in $\{0, 1, \#\}^*$, then C is not quite $E(\widetilde{\mathcal{B}})$, but instead is the union of $E(\widetilde{\mathcal{B}})$ and the set D of words in $\{0, 1, \#\}^*$ which are not the encoding of an $\mathcal{S}$-structure. However, since D is easily (deterministic polynomial-time) recognizable, it is clear that $C \in \mathcal{NP}$ iff $E(\widetilde{\mathcal{B}}) \in \mathcal{NP}$, and so there is no problem. □

It is very interesting that Theorem 18 is a statement of pure logic that seems on the surface to have nothing to do with automata theory. However, its proof is heavily dependent on automata theory.

THEOREM 19. *Let* $\mathcal{A} = \mathrm{Mod}_\omega \exists U \forall x \exists ! y(Qxy \wedge Uy)$. *Then the class of generalized spectra is closed under complement iff the complement of the* $\{Q\}$*-spectrum* $\mathcal{A}$ *is a* $\{Q\}$*-spectrum.*

PROOF. We will show that HIT $\propto E(\mathcal{A})$. Since $E(\mathcal{A}) \in \mathcal{NP}$ by Theorem 6(1), it follows that $E(\mathcal{A})$ is $\mathcal{NP}$-complete. The proof can then be completed as in Theorem 18.

Assume that e is an encoding of the family $\{A_1, \cdots, A_n\}$ of certain subsets of $\{s_1, \cdots, s_r\}$. We can assume that $n \geqslant r$ by repeating the set A_n as often as necessary. Define a finite $\{Q\}$structure $\mathfrak{A}_e$ with $|\mathfrak{A}_e| = \{1, \cdots, n\}$ such that $\langle i, j\rangle \in Q^{\mathfrak{A}_e}$ iff $s_j \in A_i$, for each i and j. Let f be a function which (in general) maps e onto the encoding of $\mathfrak{A}_e$ (and which maps nonencodings onto a fixed nonencoding). It is straightforward to check that $e \in$ HIT iff $f(e) \in E(\mathcal{A})$, and that $f \in \Pi$; therefore, HIT $\propto E(\mathcal{A})$. □

Note that $\mathcal{A}$ of Theorem 19 is a monadic $\{Q\}$-spectrum, that is, a $\{Q\}$-spectrum in which all of the "extra" predicate symbols are unary (in this case, there is only one extra predicate symbol, and it is unary). It is well known that if $\mathcal{S}$ is a set of unary predicate symbols, and $\mathcal{B}$ is a monadic $\mathcal{S}$-spectrum (that is, all predicate symbols, "given" and "extra," are unary), then there is a first-order $\mathcal{S}$-sentence σ such that $\mathcal{B} = \mathrm{Mod}_\omega \sigma$. Hence $E(\mathcal{B}) \in \mathcal{P}$, as in the proof of Theorem 6. So Theorem 19 is a best possible result (short of resolving the generalized Asser problem). We remark that the author proved the following result about monadic generalized spectra in his doctoral dissertation [9].

THEOREM 20. *Let* $\mathcal{A}$ *be the class of nonconnected finite* $\{Q\}$*-structures, where* Q *is a binary predicate symbol (a finite* $\{Q\}$*-structure* $\langle A; R\rangle$ *is connected if, for each a, b in* A*, there is a finite sequence* $a_1, \cdots, a_n$ *of points in* A *such that* $a = a_1, b = a_n$*, and either* Ra_ia_{i+1} *or* $Ra_{i+1}a_i$*, for* $1 \leqslant i < n$*). Then* $\mathcal{A}$ *is a monadic* $\{Q\}$*-spectrum, but* $\widetilde{\mathcal{A}}$ *is not a monadic* $\{Q\}$*-spectrum. In particular, the class of monadic generalized spectra is not closed under complement.*

We will now produce a "universal" NP set and a "universal" NP_1 set. Each will be complete. The technique is similar to that of Book [4], who also shows the existence of an NP_1-complete set.

Some preliminary remarks are required. If M is a one-tape nondeterministic Turing machine that operates in time T, then it is easy to see that there is a constant c and a one-tape nondeterministic Turing machine M' that recognizes A_M in time cT, such that the range of the function δ for M' (which gives the table of transitions for M') contains only sets with at most two members (intuitively, M' has at most two options per move). Whenever there are two options then by some convention we label one the first option and the other the second option.

We momentarily restrict our attention to a subclass $\mathcal{M}$ of the class of those one-tape nondeterministic Turing machines that have at most two options per move, by making natural restrictions so that $\mathcal{M}$ will be countable: We require that the sets K (of states) and Γ (of tape symbols) be subsets of ω; it is also convenient to require that the set Σ of input symbols be $\{0, 1\}$, and that each machine in $\mathcal{M}$ recognize a set of (binary representations of) positive integers. We assign Gödel numbers to machines in the class $\mathcal{M}$ in such a way that a tape head can sweep through the encoding of the Gödel number to find out how to simulate the machine with that Gödel number on a given step. One such way involves essentially letting the Gödel number be the concatenation of the entries of the table of transitions, with the # sign used as a separator. Each tape symbol and state is encoded by a string in $\{0, 1\}^*$. For details, see [13, pp. 102–103]. Denote by T_i the machine with Gödel number i.

We now define a ternary relation V, which holds for certain triples $\langle i, s, n\rangle$ with i and n positive integers, and s in $\{0, 1\}^*$. For $V(i, s, n)$ to hold, it is first necessary for i to be the Gödel number of a machine T_i. Simulate the action of T_i on the input n, in the following way: If on the kth step in the simulation, there is an option, then take the first (second) option if the kth digit in s is a 0 (1); if s is not of length at least k, then halt and reject. Then $V(i, s, n)$ holds iff the number n is accepted in this simulation.

Let t be any standard one-one map from $(Z^+)^3$ onto Z^+ such that $t \in \Pi \cap \Pi_1$ and $t^{-1} \in \Pi \cap \Pi_1$, and such that each of a, b, and c is bounded by $t(a, b, c)$. We can now define two sets of positive integers which are "universal" in the usual sense with respect to $NP(NP_1)$ sets.

$$\text{UNIV} = \{t(i, a, n): i, a, n \in Z^+ \text{ and } \exists s(\text{len}(s) = \text{len}(a) \text{ and } V(i, s, n))\},$$

$$\text{UNIV}_1 = \{t(i, a, n): i, a, n \in Z^+ \text{ and } \exists s(\text{len}(s) = a \text{ and } V(i, s, n))\}.$$

THEOREM 21. (1) *UNIV is NP-complete.* (2) *$UNIV_1$ is NP_1-complete.*

PROOF. UNIV $\in NP$. For, we can define a multi-tape nondeterministic machine M which, given $t(i, a, n)$ as input, finds i, a, and n, guesses s in $\{0, 1\}^*$ such that $\mathrm{len}(s) = \mathrm{len}(a)$, and then does the obvious simulation. The point is that a is so large that the time of simulation is (except for bookkeeping) equal to the length of a, which is bounded by the length of $t(i, a, n)$; hence, UNIV $\in NP$. Similarly, $\mathrm{UNIV}_1 \in NP_1$, since the time of simulation is roughly given by a, which is roughly 2^l, where l is the length of a.

Assume that $B \in NP$; we want to show that $B \propto$ UNIV. By the usual encoding arguments, we can assume that $B \subseteq Z^+$. Find i and k such that T_i recognizes B in time $l \longmapsto l^k$. Then $n \in B$ iff $t(i, a, n) \in$ UNIV, where a is a string of $(\mathrm{len}(n))^k$ tallies. Clearly the function $n \longmapsto t(i, a, n)$ is in Π. So UNIV is NP-complete.

Now assume that the set B of positive integers is in NP_1. Find i and k such that T_i recognizes B, and T_i accepts each n in B within n^k steps. Then $n \in B$ iff $t(i, n^k, n) \in \mathrm{UNIV}_1$. So UNIV_1 is NP_1-complete. □

We are especially interested in part 2 of Theorem 21, which gives us a particular spectrum whose complement is a spectrum iff the complement of every spectrum is a spectrum. We record this in Theorem 22.

THEOREM 22. *The class of spectra is closed under complement iff the complement of the spectrum $UNIV_1$ is a spectrum.*

The proof is immediate. □

COROLLARY 23. *There is an NP_1-complete set in BIN. Thus, this is an example of a spectrum A in BIN such that $\widetilde{A}$ is a spectrum iff the complement of every spectrum is a spectrum.*

PROOF. Find k from Theorem 8 such that $A = \{n^k : n \in \mathrm{UNIV}_1\}$ is in BIN. We remark that a simple analysis shows that in this case, k can be taken to be 5. Then $n \in \mathrm{UNIV}_1$ iff $n^k \in A$, for each n. Hence $\mathrm{UNIV}_1 \propto_1 A$, and so A is NP_1-complete. □

8. A Savitch-like result. Savitch [20] showed that any set that can be recognized nondeterministically in tape S can be recognized deterministically in tape S^2. If such a theorem were true for time bounds—for example, if there were a constant k such that any set that can be recognized nondeterministically in time T can be recognized deterministically in time T^k—then, of course, it would follow that $NP = P$ and $NP_1 = P_1$. It is quite surprising that this strong condition we are discussing is essentially equivalent to the apparently weaker condition that $NP = P$.

We will prove this in Theorem 24. Then we will generalize the result in various ways, and conclude by an analogy with Post's problem.

THEOREM 24. *The following two statements are equivalent*:

1. $NP = P$.

2. *There exists a constant* k *such that, for every countable function* T *with* $T(l) \geqslant l + 1$ *for each* l *and for every language* A *which is recognized by a nondeterministic one-tape Turing machine in time* T, *the language* A *is recognized by a deterministic one-tape Turing machine in time* T^k.

PROOF. $2 \Rightarrow 1$: This is immediate, since $l \longmapsto l^k$ is countable for each k.

$1 \Rightarrow 2$: It is sufficient to show that $1 \Rightarrow 2'$, where $2'$ is obtained from 2 by replacing both occurrences of "language A" by "set $A \subseteq Z^+$." This is because of simple interrelationships between machines M which recognize a language A and machines M' which recognize an encoding $A' \subseteq Z^+$ of A. The details are straightforward and nonunique, and are left to the reader.

Let $R = \{\bar{i} \# \bar{a} \# \bar{n}$: if $\bar{i}, \bar{a}$, and $\bar{n}$ are the binary representations of the positive integers i, a, and n, then $t(i, a, n) \in \text{UNIV}\}$. Then $R \in NP$, and so by hypothesis (and by Theorem 2), there is a constant k' and a one-tape deterministic machine M_1 which recognizes R in time $l \longmapsto l^{k'}$. We can assume that $k' \geqslant 2$. Let $k = 2k'$.

Assume that A is recognized by a nondeterministic one-tape machine in time T, where T is countable and $T(l) \geqslant l + 1$ for each l. Then as we observed earlier, there is a constant c_1 and a machine T_{i_0} (with at most two options per move) which recognizes A in time $c_1 T$. Since T is countable, it is easy to see that $c_1 T$ is countable. Hence there is a constant c_2 and a deterministic two-tape machine M_2 which, for each l and each input w of length l on the first tape, prints at least $c_1 T(l)$ tallies on its second tape in at most $c_2 T(l)$ steps.

We will now describe a 3-tape nondeterministic machine M which recognizes A. Given input n of length l on its first tape, M simulates M_2 to print a string w of at least $c_1 T(l)$ tallies on its second tape in at most $c_2 T(l)$ steps. Then M prints $\bar{i}_0 \# w \# \bar{n}$ on its third tape in $\text{len}(i_0) + \text{len}(w) + l + 2$ steps. Now M simulates M_1 with $\bar{i}_0 \# w \# \bar{n}$ as input. This takes at most $(\text{len}(i_0) + \text{len}(w) + l + 2)^{k'}$ steps. Since $T(l) \geqslant l + 1$, since clearly $\text{len}(w) \leqslant c_2 T(l)$, and since $\text{len}(i_0) + 2$ is a constant, the total number of steps required is bounded by $((c_2 + 2)T(l))^{k'}$ for sufficiently large l. Clearly, M recognizes A. By Theorem 2, the set A is recognized by a one-tape deterministic machine in time $((c_2 + 2)T)^k$. Hence, by Theorem 1, A is recognized by a one-tape deterministic machine in time T^k. □

By very similar proofs, we can demonstrate the following two results.

THEOREM 25. *The following two statements are equivalent*:

1. $NP_1 = P_1$.

2. *There exists a constant* k *such that, for every countable function* T *with* $T(l) \geqslant 2^l$ *for each* l *and for every language* A *which is recognized by a nondeterministic one-tape Turing machine in time* T, *the language* A *is recognized by a deterministic one-tape Turing machine in time* T^k.

THEOREM 26. *The following two statements are equivalent.*

1. NP (NP_1) *is closed under complement.*

2. *There exists a constant* k *such that, for every countable function* T *with* $T(l) \geqslant l + 1$ $(T(l) \geqslant 2^l)$ *for each* l *and for every language* A *which is recognized by a nondeterministic one-tape Turing machine in time* T, *the language* $\widetilde{A}$ *is recognized by a nondeterministic one-tape Turing machine in time* T^k.

We conclude this section by an analogy with Post's problem. Definitions and notation are from Rogers [19]. Post's problem asks whether there is an r.e. set C which is not Turing-equivalent to either $\varnothing$ or to the halting problem set K.

Let $\{W_x^B: x \in Z^+\}$ be an effective listing of all sets of natural numbers which are r.e. in B. As Rogers notes, if A and B are r.e., then the assertion that A is not Turing-reducible to B is equivalent to $\forall x(\widetilde{A} \neq W_x^B)$, or equivalently, $\forall x \exists y(y \in A$ iff $y \in W_x^B)$. If $(\exists$ recursive $f)$ $(\forall x)(f(x) \in A$ iff $f(x) \in W_x^B)$, then we say that A *is constructively nonrecursive in* B.

Many attempts to solve Post's problem failed, because investigators tried to find some r.e. set C such that C is constructively nonrecursive in $\varnothing$ and such that K is constructively nonrecursive in C. Rogers shows that if A and B are r.e., and if A is constructively nonrecursive in B, then B is recursive. Hence, any such attempt must fail.

In an analogous way, one might wonder whether it is possible that $NP = P$, but that all attempts to prove this have failed because investigators have been searching for some recursive function f which maps the index i of each nondeterministic Turing machine into an index $f(i)$ of a deterministic machine which recognizes the same set, such that if the machine with index i operates in polynomial time, then so does the machine with index $f(i)$. We will now show that if $NP = P$, then there is such a recursive function f, as long as we restrict ourselves to machines that operate in a given polynomial time bound, such as machines that operate in time $l \longmapsto l^r$ for fixed r.

For each r, let T_i^r be a two-tape nondeterministic machine which, given input n on its first tape, simulates the action of T_i on n for at most $(\operatorname{len}(n))^r$ steps, by using its second tape as a clock. If in the simulation T_i has not accepted within $(\operatorname{len}(n))^r$ steps, then T_i^r halts and rejects.

THEOREM 27. *The following two statements are equivalent:*

1. $NP = P$.

2. *There exists a constant* k *and a function* f *in* Π *such that, for each Gödel number* i *and each positive integer* r, *the machine* $T_{f(i,r)}$ *is a one-tape deterministic Turing machine which operates in time* $l \mapsto l^{kr}$, *and which recognizes the same set as* T_i^r. *Hence, if* T_i *operates (nondeterministically) in time* $l \mapsto l^r$, *then* $T_{f(i,r)}$ *recognizes the same set as* T_i.

PROOF. This is clear from the proof of Theorem 24. □

9. A counterexample. We will show that there is a spectrum S in BIN such that $\{n: 2^n \in S\}$ is not a spectrum. By way of contrast, it is easy to see, because of Theorem 6(2), that for each spectrum S and each polynomial p with rational coefficients the set $\{n: p(n) \in S\}$ is a spectrum. The fact that there is a spectrum S such that $\{n: 2^n \in S\}$ is not a spectrum is extremely closely related, both in content and method, to Bennett's results on higher-order spectra [2], although he did not specifically state or prove this result. If we analyze Bennett's proof, then we see that he essentially proved that there is a spectrum S and a positive integer k such that $\{n: 2^{n^k} \in S\}$ is not a spectrum.

We will also show that our techniques give a new proof of a result of Book [3] that $NP \neq NP_1$.

LEMMA 28. *Let* A *be a spectrum. Then, for some constant* k, *the set* A *is recognized by a one-tape deterministic Turing machine in tape* $l \mapsto 2^{kl}$.

PROOF. Assume first that $A = \{n: \langle n \rangle \models \exists Q\sigma\}$, where Q is a binary predicate symbol. Define a one-tape deterministic machine M which, given input n, systematically prints all possible strings in $\{0, 1\}^*$ of length n^2, and tests them one by one to see if the binary relation R on n which the string represents in the natural way has the property that $\langle n; R \rangle \models \sigma$. M accepts n iff it finds some such string. If $\text{len}(n) = l$, then $n^2 < 2^{2l}$. Hence M can be arranged to operate in tape $l \mapsto 2^{3l}$. Similarly, for each spectrum S there is a constant k such that A is recognizable in tape $l \mapsto 2^{kl}$. □

LEMMA 29. *There is a set* $A \subseteq Z^+$ *which is recognized by a one-tape deterministic Turing machine in tape* $l \mapsto 2^{l^2}$, *which is not a spectrum.*

PROOF. This follows from Theorem 5 and Lemma 28, since it is easy to see that $l \mapsto 2^{l^2}$ is constructible and that $\lim \inf_{l \to \infty} 2^{kl}/2^{l^2} = 0$ for each k.

LEMMA 30. *Assume that* $A \subseteq Z^+$ *is recognized by a one-tape deterministic Turing machine in tape* $l \mapsto 2^{l^2}$. *Then there is a set* B *in* E^2_* *such that* $A = \{n: 2^{2^n} \in B\}$.

PROOF. Let M be a one-tape deterministic Turing machine that recognizes A in tape $l \longmapsto 2^{l^2}$. Let M_1 be a one-tape deterministic Turing machine which operates as follows: Given input m, the machine M_1 tests to see if there is a positive integer n such that $m = 2^{2^n}$. If not, then M_1 rejects. If so, then M_1 simulates M on input n. Now M_1 can be designed to be a linear-bounded automaton. This is because $\text{len}(2^{2^n}) = 2^n + 1$, which is bigger than $2^{2^{l-1}}$ (where $l = \text{len}(n)$), which is bigger than 2^{l^2} for sufficiently large l. So, by Theorem 3, the set B which M_1 recognizes is in E_*^2. Clearly $A = \{n: 2^{2^n} \in B\}$. □

THEOREM 31. *There is a spectrum* S *such that* $\{n: 2^n \in S\}$ *is not a spectrum.*

PROOF. Find A from Lemma 29 and B from Lemma 30 such that A is not a spectrum, $B \in E_*^2$, and $A = \{n: 2^{2^n} \in B\}$. By Theorem 7, we know that B is a spectrum. Let $C = \{n: 2^n \in B\}$. Then $A = \{n: 2^n \in C\}$. Assume that it is always true that whenever S is a spectrum, then $\{n: 2^n \in S\}$ is a spectrum. Then C is a spectrum (since B is), and so A is a spectrum (since C is). But this is a contradiction. □

COROLLARY 32. *There is a spectrum* T *in BIN such that* $\{n: 2^n \in T\}$ *is not a spectrum.*

PROOF. Find S from Theorem 31 such that $D = \{n: 2^n \in S\}$ is not a spectrum. Find a positive integer k from Theorem 8 such that $T = \{n^k: n \in S\}$ is in BIN. Let $E = \{n: 2^n \in T\}$. Then $n \in D$ iff $kn \in E$, for each positive integer n; for, $n \in D$ iff $2^n \in S$ iff $2^{kn} \in T$ iff $kn \in E$. If E were a spectrum, then E would be in NP_1, and so clearly D would be in NP_1. Hence D would be a spectrum, a contradiction. □

We close this section with some further observations. Theorem 13 of §6 could just as well have been stated as follows:

(4) *Assume* $A \subseteq Z^+$. *If* $A \in NP$, *then* $\{n: 2^n \in A\}$ *is in* NP_1.

We remark that we can use the technique of the proof of Lemma 30 to show that (4) has a converse:

THEOREM 33. *Assume* $B \subseteq Z^+$. *Then* $B \in NP_1$ *iff there is* A *in* NP *such that* $B = \{n: 2^n \in A\}$.

Because of Theorem 6(2), we know that Theorem 31 can be restated as follows:

(5) *There is a set* A *in* NP_1 *of positive integers such that* $\{n: 2^n \in A\}$ *is not in* NP_1.

Similarly, we can prove the following:

THEOREM 34. *There is a set A in NP of positive integers such that $\{n: 2^n \in A\}$ is not in NP.*

Finally, we observe that (4) and Theorem 34 combine to give us a theorem of Book:

THEOREM 35 (BOOK [3]). $NP \subsetneq NP_1$.

PROOF. From Theorem 34, find a set A in NP of positive integers such that $B = \{n: 2^n \in A\}$ is not in NP. By (4), we know that $B \in NP_1$. So $NP \neq NP_1$. Of course, $NP \subseteq NP_1$. □

Book's proof depends on a fairly difficult result of Cook [8]. No simple diagonalization argument seems capable of proving Theorem 35 directly, because we are dealing with nondeterministic, rather than deterministic, time-complexity classes. However, a simple diagonalization argument does show that $P \subsetneq P_1$.

10. A real-time recognizable NP-complete set. We conclude by exhibiting an NP-complete set REAL which is recognized by a nondeterministic two-tape machine in real time. The existence of such a set is not new: Hunt [14] shows the existence of an NP-complete set which is recognizable nondeterministically in linear time, and Book and Greibach [5] prove that every set recognizable nondeterministically in linear time is recognizable by a nondeterministic two-tape Turing machine in real time. However, our set is produced directly, and is fairly simple. The existence of such a set is a best-possible result, since Rabin and Scott [17] show that every set which is recognized by a one-tape nondeterministic Turing machine in real time is recognized by a one-tape deterministic Turing machine in real time.

Let REAL $= \{a_1 \# a_2 \# \cdots \# a_{2r}: r \in Z^+; a_i \in \{0, 1, 2\}^*$ for each i; $\text{len}(a_i) = \text{len}(a_j)$ for each i, j; and there exists b in $\{0, 1\}^*$ such that $\text{len}(b) = \text{len}(a_i)$ for each i, and such that for each odd i there exists k such that the kth member of the string b and the kth member of the string a_i are the same$\}$.

THEOREM 36. *REAL is an NP-complete set which is recognized by a two-tape nondeterministic Turing machine in real time.*

PROOF. Let M be a two-tape nondeterministic Turing machine which works as follows: As a_1 is being read on the first, or input tape, M nondeterministically prints some b in $\{0, 1\}^*$ on the second tape, such that $\text{len}(b) = \text{len}(a_1)$; meanwhile, M checks to make sure that, for some k, the kth digit of b

is the same as the kth digit of a_1. When M reads # on the input tape and starts reading a_2, the second tape head runs back over b on the second tape and uses the length of b to measure the length of a_2. If $\mathrm{len}(a_2) \neq \mathrm{len}(b)$, then M halts and rejects. If $\mathrm{len}(a_2) = \mathrm{len}(b)$, then the tape heads are in a position to compare b and a_3 digit by digit. M continues in the obvious way. Clearly, M recognizes REAL in real time.

SAT $\propto$ REAL: Let θ be a conjunctive normal form expression, with clauses $C_1, \cdots, C_r$, and propositional letters $A_1, \cdots, A_n$. (If $\theta = \bigwedge_i \bigvee_j B_{ij}$, then each $\bigvee_j B_{ij}$ is a clause.) We can assume that no clause C_i contains both A_k and $\sim A_k$ for any k, or else that clause can be eliminated. Let β_θ be the expression $a_1 \# a_2 \# \cdots \# a_{2r}$, where each a_i is of length n, where if i is even, then a_i is a string of tallies, and where if $i = 2s - 1$ is odd, then for each k $(1 \leqslant k \leqslant n)$, the kth digit of a_i is as follows:

$$\begin{cases} 0, & \text{if } \sim A_k \text{ appears in the } s\text{th clause,} \\ 1, & \text{if } A_k \text{ appears in the } s\text{th clause,} \\ 2, & \text{otherwise.} \end{cases}$$

For any reasonable encoding e, there exists a constant c such that if the encoding $e(\theta)$ of θ is of length l, then $l \geqslant c \cdot \max(r, n)$. Now β_θ has length $2rn + 2r - 1$, which is dominated by $2l^2/c + 2l/c - 1$. So if f is the function which (in general) maps $e(\theta)$ onto β_θ (and which maps strings not of the form $e(\theta)$ onto a fixed string not in REAL), then it is easy to see that $f \in \Pi$ (we are assuming that $\{e(\theta)$: θ is a formula in conjunctive normal form$\}$ is in P, which is also true for any reasonable encoding e). Most importantly, it is clear that θ is satisfiable iff $\beta_\theta \in$ REAL. Hence, SAT $\propto$ REAL. □

Bibliography

1. G. Asser, *Das Repräsentantenproblem im Prädikatenkalkül der ersten Stufe mit Identität*, Z. Math. Logik Grundlagen Math. 1 (1955), 252–263. MR 17, 1038.

2. J. H. Bennett, *On spectra*, Doctoral Dissertation, Princeton University, Princeton, N. J., 1962.

3. R. V. Book, *On languages accepted in polynomial time*, SIAM J. Comput. 1 (1972), 281–287.

4. ———, *Comparing complexity classes* (submitted for publication).

5. R. V. Book and S. Greibach, *Quasi-realtime languages*, Math. Systems Theory 4 (1970), 97–111. MR 43 #1772.

6. A. Cobham, *The intrinsic computational difficulty of functions*, Logic, Methodology, and Philos. (Proc. 1964 Internat. Congress), North-Holland, Amsterdam, 1965, pp. 24–30. MR 34 #7376.

7. S. Cook, *The complexity of theorem-proving procedures*, Conference Record of Third ACM Sympos. on Theory of Computing, 1970, pp. 151–158.

8. S. Cook, *A hierarchy for nondeterministic time complexity*, Proc. Fourth ACM Sympos. on Theory of Computing, 1972, pp. 187–192.

9. R. Fagin, *Contributions to the model theory of finite structures*, Doctoral Dissertation, University of California, Berkeley, Calif., 1973.

10. A. Grzegorczyk, *Some classes of recursive functions*, Rozprawy Mat. **4** (1953), pp. 1–45. MR **15**, 667.

11. J. Hartmanis, P. M. Lewis II and R. E. Stearns, *Hierarchies of memory limited computations*, IEEE Conference Record on Switching Circuit Theory and Logical Design, Ann Arbor, Mich., 1965, pp. 179–190.

12. J. Hartmanis and R. E. Stearns, *On the computational complexity of algorithms*, Trans. Amer. Math. Soc. **117** (1965), 285–306. MR **30** #1040.

13. J. E. Hopcroft and J. D. Ullman, *Formal languages and their relation to automata*, Addison-Wesley, Reading, Mass., 1969. MR **38** #5533.

14. H. B. Hunt III, *On the time and tape complexity of languages*, Technical Report 73–156, Cornell Univ., Ithaca, N. Y., Jan. 1973.

15. N. D. Jones and A. L. Selman, *Turing machines and the spectra of first-order formulas with equality*, Proc. Fourth ACM Sympos. on Theory of Computing, 1972, pp. 157–167.

16. R. M. Karp, *Reducibility among combinatorial problems*, Technical Report 3, University of California, Berkeley, April 1972; Also in *Complexity of computer computations* (ed. R. E. Miller et al.), Plenum Press, New York, 1972.

17. M. O. Rabin and D. Scott, *Finite automata and their decision problems*, IBM J. Res. Develop. 3 (1959), pp. 114–125; Also in *Sequential machines: selected papers* (ed. E. F. Moore), Addison-Wesley, Reading, Mass., 1964, pp. 63–91. MR **21** # 2559.

18. R. W. Ritchie, *Classes of predictably computable functions*, Trans. Amer. Math. Soc. **106** (1963), 139–173. MR **28** #2045.

19. H. Rogers, *Theory of recursive functions and effective computability*, McGraw-Hill, New York, 1967. MR **37** #61.

20. W. J. Savitch, *Relationships between nondeterministic and deterministic tape complexities*, J. Comput. Systems Sci. **4** (1970), 177–192. MR **42** #1605.

21. H. Scholz, J. Symbolic Logic **17** (1952), 160.

22. J. R. Shoenfield, *Mathematical logic*, Addison-Wesley, Reading, Mass., 1967. MR **37** #1224.

23. A. Tarski, *Contributions to the theory of mdoels.* I, II, Nederl. Akad. Wetensch. Proc. Ser. A. **57** = Indag. Math. **16** (1954), 572–588. MR **16**, 554.

24. A. M. Turing, *On computable numbers with an application to the Entscheidungsproblem*, Proc. London Math Soc. (2) **42** (1936), 230–265; correction, ibid. (2) **43** (1937) 544–546.

T. J. WATSON RESEARCH CENTER, IBM

SIAM-AMS Proceedings
Volume 7
1974

On k-Tape Versus $(k-1)$-Tape Real Time Computation

S. O. Aanderaa

Abstract. We shall construct for each natural number k a language L_k which can be recognized in real time by a k-tape Turing machine, but which cannot be recognized in real time by any $(k-1)$-tape Turing machine. In fact, it turns out that the language L_k above can be recognized in real time by a k-pushdown-store machine. The proof is based on the notion of overlap.

Introduction. Although a Turing machine with one tape can compute all functions which can be computed by a Turing machine with several tapes, it seems reasonable to expect that adding tapes to a multi-tape Turing machine will speed up the computation in some cases. Adding new tapes beyond two tapes, however, does not speed up the computation very much, as shown by Hennie and Stearns. They proved in a paper [9] published in 1966 that if a given function requires computation time T for a k-tape realization, then it requires at most computation time $T \log T$ for a two-tape realization. The first negative result was obtained as early as 1963 by Rabin [11]. He proved that there exists a set T_2 which is real time definable by a two-tape Turing machine, but not by any one-tape Turing machine.

We shall improve Rabin's result and show that, for any $k \geqslant 2$, there exists a language L_k which can be recognized in real time by a k-tape Turing machine but not by a $(k-1)$-tape Turing machine. Although we cannot prove the converse of Hennie and Stearns' result, our method of proof tends to support the hypothesis that Hennie and Stearns' result cannot be improved. Our proof is longer and more complicated than Rabin's, and uses the notion of overlap. The notion of overlap appeared originally in Cook [2] and in Cook and Aanderaa [3].

AMS (MOS) subject classifications (1970). Primary 68A20; Secondary 68A25, 02F1, 02F15.

Key words and phrases. Real time computation, k-tape turing machines, overlap, computational complexity.

The method of calculating overlap has been improved by Fischer, Meyer and Paterson [10], by introducing the notion of average overlap, which we shall use in this paper.

We shall now give an informal discussion of k-tape real time Turing machines. A k-tape real time Turing machine has k working tapes. It may also have an input tape, which is a one-way read-only tape (or it may have instead an input terminal), and it may also have an output tape which is a one-way write-only tape (or it may have instead an output terminal). A k-tape Turing machine with output tape and no input tape can be used to generate a sequence (see [1] and [4]). Off-line Turing machines in the sense of Hennie are supposed to have the input written on one of the working tapes, and such machines do not have an input tape (see Hennie [8]).

We shall here in this paper consider only Turing machines which do have input tapes. With respect to output tapes we have a choice. We have decided to treat our k-tape real time Turing machines as acceptors, so that a Turing machine accepts or rejects an input depending on its final state when all the input is read. We could as well have worked with functions which map initial segments of the input tape into the output alphabet $\{0, 1\}$, so that a Turing machine writes 1 on the output tape if the initial segment of the input is accepted, and 0 otherwise. See for instance [3, pp. 296–297].

1. Real time Turing machines. We shall here give a formal definition of a k-tape real time Turing machine. Our definition is a modification of Rosenberg's definition [12, pp. 646–647].

For finite sets A let $\#(A)$ denote the cardinality of A. Let $A \times B$ denote the cross-product of A and B, i.e., $A \times B = \{(a, b) | a \in A \text{ and } b \in B\}$. The sets $[A]^i$, $i = 1, 2, 3$, are defined recursively by

$$[A]^1 = A, \qquad [A]^{i+1} = A \times [A]^i.$$

Let Γ be a finite alphabet. Then Γ^* is the set of words over Γ. Γ^n is the set of words over Γ of length n. Let w be a word. Then $|w| =$ length of w.

DEFINITION. A *real time k-tape on line Turing machine* (abbreviated k-RTTM) is an 8-tuple $(K, \Gamma, \Sigma, \sigma, \lambda, q_0, b, F)$ where:

1. K is a nonempty finite set of states.
2. Γ is an alphabet (of input symbols).
3. Σ is an alphabet (of working symbols).
4. σ is the state-transition function which maps $K \times \Gamma \times [\Sigma]^k$ into K.
5. λ is the action function which maps $K \times \Gamma \times [\Sigma]^k$ into $[\{-1, 0, 1\}]^k \times [\Sigma]^k$.

6. q_0 (the initial state) is a member of K.
7. b (the blank symbol) is a member of Σ.

An *instantaneous description* (i.d.) may be denoted by the $(k+2)$-tuple

(1)
$$(q, B, x_1my_2, \cdots, x_kmy_k)$$

where $q \in K$, $B \in \Gamma^*$, $x_i \in \Sigma^*\Sigma$, $y_i \in \Sigma^*\Sigma$ and $m \notin \Sigma$. ($\Sigma^*\Sigma$ is the set of nonempty words over Σ.) It turns out to be more convenient to represent an i.d. by the expression

(2)
$$\langle q, A, t, \alpha, x_1, x_2, \cdots, x_k\rangle$$

where $q \in K$; and $\alpha = \langle \alpha_1, \alpha_2, \cdots, \alpha_k\rangle$ and each α_i is a mapping of $Z = \{\cdots, -1, 0, 1, 2, \cdots\}$ into Σ such that $\{x \mid x \in Z \;\&\; \alpha_i(x) \neq b\}$ is a finite set, $A \in \Gamma^*$, and $0 \leqslant t \leqslant |A|$. Moreover $x_i \in Z$, and x_i in (2) means that square number x_i on the tape T_i is scanned. To each i.d. represented in the form (2), there corresponds a *display* $\langle q, a, s_1, s_2, \cdots, s_k\rangle$ where $q \in K$, $s_i \in \Sigma$, $a \in \Gamma$ and $s_i = \alpha_i(x_i)$ and a is the symbol in position $t+1$ of the string A (the display is not defined if t exceeds the length of A). Let the display of $\langle q, A, t, \alpha, x_1, \cdots, x_k\rangle$ be $d = \langle q, a, s_1, \cdots, s_k\rangle$, let

$$\lambda(d) = (\epsilon_1, \epsilon_2, \cdots, \epsilon_k, s'_1, s'_2, \cdots, s'_k),$$

and let

$$\alpha = \langle \alpha_1, \alpha_2, \cdots, \alpha_k\rangle \quad \text{and} \quad \alpha' = \langle \alpha'_1, \alpha'_2, \cdots, \alpha'_k\rangle;$$

then

$$\langle q, A, t, \alpha, x_1, \cdots, x_k\rangle \vdash \langle q', A', t', \alpha', x'_1, \cdots, x'_k\rangle$$

if the following conditions are satisfied:

1. $q' = \sigma(d)$ and $A' = A$, $t' = t+1$, $t \leqslant |A|$.
2. $x'_j = x_j + \epsilon_j$ for all $j = 1, 2, \cdots, k$.
3. For each $j = 1, 2, \cdots, k$, we have that

$$\alpha'_j(x) = \alpha_j(x), \text{ for all } x \in Z - \{x_j\} \quad \text{and} \quad \alpha'_j(x_j) = s'_j.$$

We define $\vdash^*$ to be the transitive closure of $\vdash$, i.e. $D \vdash^* D'$, if there exists i.d.'s $D_0, D_1, \cdots, D_m$ such that $D_0 = D$ and $D_m = D'$ and $D_j \vdash D_{j+1}$ for all $j = 0, 1, 2, \cdots, m-1$.

An i.d. $D = \langle q, A, t, \alpha, x_1, x_2, \cdots, x_k\rangle$ where $\alpha = \langle \alpha_1, \alpha_2, \cdots, \alpha_k\rangle$ is an *initial* i.d. with input W if the following conditions are satisfied:

1. $q = q_0$, $A = W$, $t = 0$.
2. $\alpha_i(j) = b$ for all $i = 1, 2, \cdots, k$ and all $j \in Z$.
3. $x_1 = x_2 = \cdots = x_k = 0$.

A computation of the word $W \in \Gamma^*$ is a sequence of i.d.'s $D_0, D_1, \cdots, D_N$ where D_0 is an initial i.d. with input W, and $N = |W|$, and $D_i \vdash D_{i+1}$ for all $i = 0, 1, 2, \cdots, N-1$. D_i $(1 \leqslant i \leqslant N)$ is the i.d. just after the machine has acted on the ith input symbol.

A work $W \in \Gamma^*$ is *accepted* by $M = (K, \Gamma, \Sigma, \sigma, \lambda, q_0, b, F)$ if there exists a computation $D_0, D_1, \cdots, D_N$ of W such that $D_N = \langle q, W, N, \alpha, x_1, \cdots, x_k \rangle$ and $q \in F$. The set of words accepted by M is denoted by $T(M)$. M is said to recognize the language L, if $L = T(M)$.

We shall say that a k-tape real time Turing machine is a k-pushdown tape machine if each of the heads on the working tapes writes b (the blank symbol) when the head moves to the left, i.e., for each i and each display $d \in K \times \Gamma \times [\Sigma]^k$ we have that if $\lambda(d) = (\epsilon_1, \epsilon_2, \cdots, \epsilon_i, \cdots, \epsilon_k, s_1, s_2, \cdots, s_i, \cdots, s_k)$ and if $\epsilon_i = -1$, then $s_i = b$.[1]

The result. The result of the paper may be stated as follows:

THEOREM 1. *Let k be a positive integer. Then there exists a language L_k which can be recognized in real time by a k-pushdown-tape machine but which cannot be recognized by a $(k-1)$-tape real time Turing machine.*

Let C_k and C'_k denote the class of languages which can be recognized by a k-tape real time Turing machine and k-pushdown-tape real time machine, respectively. Then we may restate the theorem as follows.

THEOREM 1′. $C'_k - C_{k-1} \neq \emptyset$ *for all* $k = 1, 2, \cdots$.

As corollaries we obtain

COROLLARY 1. $C_{k-1} \neq C_k$.

COROLLARY 2. $C'_{k-1} \neq C'_k$.

We would also like to add the following conjecture.

CONJECTURE. $C_k - C'_{2k-1} \neq \emptyset$.

2. The notion of overlap. Let $D = \langle q, W, N, \alpha, x_1, x_2, \cdots, x_k \rangle$, and let $\langle q, a, s_1, s_2, \cdots, s_k \rangle$ be the display of D. The *extended display* of D is the $(2k+3)$-tuple $\langle q, a, s_1, s_2, \cdots, s_k, N, x_1, x_2, \cdots, x_k \rangle$.

Let $W \in \Gamma^*$, let $N = |W|$, and let $D_0, D_1, \cdots, D_N$ be a computation of W. Then an *input interval* is a sequence of consecutive integers chosen from

[1] This is not the usual definition of a pushdown machine. For example, we can print only one symbol on the store on a given move. But it is not hard to see that there is no loss in generality in using our definition.

the set $\{0, 1, 2, \cdots, N\}$. If $I = \{g, g+1, \cdots, h\}$ is an input interval then the *time interval* $I^{\tau}(W)$ (which we also write as $I^{\tau,W}$) is the sequence $\{D_g, D_{g+1}, \cdots, D_h\}$. The *display sequence* $I^d(W)$ of I is the sequence $\{d_g, d_{g+1}, \cdots, d_h\}$ where d_i is the display for D_i for all $i = g, g+1, \cdots, h$. The *extended display sequence* $I^{ed}(W)$ of I is the sequence $\{e_g, e_{g+1}, \cdots, e_h\}$ where e_i is the extended display of D_i for all $i = g, g+1, \cdots, h$.

Let $W_1 \in \Gamma^*$ and let $W_2 \in \Gamma^*$ and let $|W_1| = |W_2| = N_1$ and let I be an input interval. Then we shall say that the computation of W_1 and W_2 are *I-equivalent* if $I^{ed}(W_1) = I^{ed}(W_2)$.

Let $D_0, D_1, D_2, \cdots, D_N$ be a computation of W and let

$$D_0 = \langle q_0, W, 0, \alpha, x_{01}, x_{02}, \cdots, x_{0k}\rangle,$$

$$D_1 = \langle q^{(1)}, W, 1, \alpha^{(1)}, x_{11}, x_{12}, \cdots, x_{1k}\rangle,$$

$$D_j = \langle q^{(j)}, W, j, \alpha^{(j)}, x_{j1}, x_{j2}, \cdots, x_{jk}\rangle,$$

$$D_N = \langle q^{(N)}, W, N, \alpha^{(N)}, x_{N1}, x_{N2}, \cdots, x_{Nk}\rangle.$$

Then the ith head movement function η_i^W during the computation of a word W is defined as: $\eta_i^W(j+1) = x_{ji}$ = position of head just before reading the $(j+1)$st symbol. We shall use the expression $[i, j]$ to denote the interval $\{i, i+1, \cdots, j\}$. Let $I = [i, j]$. Then, for each choice of the word $A \in \Gamma^N$, we assign a subword $A(I)$ of A to the interval I; namely the subword of A consisting of symbols in position $i, i+1, \cdots, j$.

Let the k working tapes be denoted by $T_1, T_2, \cdots, T_k$. Then we use the pair $\langle i, s\rangle$ to denote square number s on tape T_i. We shall use the expression $S^A(I)$ to denote the locations scanned by some head during the time interval $I^{\tau}(A)$. Two input intervals $I = [g, h]$ and $J = [g', h']$ are *adjacent* if $g' = h+1$. Let I and J be adjacent input intervals. Then the overlap caused by the input A during input intervals I and J is

$$\omega^A(I, J) = \#(S^A(I) \cap S^A(J)).$$

Let also

$$\Omega^A(I, J) = S^A(I) \cap S^A(J).$$

(Here $\#(S)$ denotes the cardinality of the set S.)

Let $I = [g, h]$. Then the length $|I|$ of I is $h + 1 - g$. The *internal overlap* caused by the input A during interval I is

$$\omega^A(I) = \max_{g \leq i < h} (\omega^A([g, i], [i+1, h]))$$

and the internal overlap coefficient is

$$\xi^A(I) = \omega^A(I)/(h + 1 - g) = \omega^A(I)/|I|.$$

Let $S \subseteq \Gamma^N$, and let $\gamma = 1/\#(S)$. Then the *average internal overlap* caused by S during interval I is

$$\omega^S(I) = \gamma \sum_{A \in S} \omega^A(I)$$

and the *average internal overlap coefficient* caused by S during interval I is

$$\xi^S(I) = \gamma \sum_{A \in S} \xi^A(I) = \omega^S(I)/|I|.$$

REMARK. The notion of overlap $\omega^A(I, J)$ was first defined and used by Cook and Aanderaa [2] and [3]. The idea of considering average overlap is due to Fischer, et al. [10]. The notion of internal overlap is a new concept.

We shall in §4 prove the following lemma.

LEMMA 1 (THE OVERLAP LEMMA). *Let M be a $(k-1)$-tape real time Turing machine with input alphabet Γ. Let N_0 and N be positive integers and let $N > 100N_0$. Let $S \subseteq \Gamma^N$. Then there exists an input interval $I = [g, h]$ such that $|I|/N_0$ is an integer and $0 \leq g \leq h \leq N$ and such that*

$$\xi^S(I) \leq (4(k-1) \log \log (N/N_0))/\log (N/N_0).$$

3. Outline of the proof. In this section, we shall try to motivate and outline the proof of the main result. We need to define a "retrieval language" L_k (see §6) over the alphabet $\Gamma = \Gamma_1 \cup \cdots \cup \Gamma_k$ where $\Gamma_i = \{a_i, b_i, b_i'\}$. With k tapes, one work tape is devoted to each Γ_i. Intuitively, with fewer than k tapes, one tape has to store and retrieve symbols from both Γ_i and Γ_j $(i \neq j)$.

It suffices to define a set $S \subseteq \Gamma^*$ for which the assumption that a real time $(k-1)$-tape machine works properly on S leads to a contradiction. S is defined (see §7) so that the density of symbols from Γ_i is greater than that from Γ_j $(i < j)$. So intuitively, if the overworked tape "concentrates" on Γ_i, it will not have enough time to run back for the Γ_j symbols; if it concentrates on Γ_j it will not have enough space to remember all the Γ_i symbols. The difficulty lies in trying to formalize this intuition since the $k-1$ tapes may be encoding information in a very clever manner.

In addition to the concept of overlap, we will need the following notions (see §5).

(i) Two i.d.'s are *z-equivalent* (i.e., equivalent up to a displacement of z squares of any tape head).

(ii) Two adjacent intervals $[g, h]$ and $[h + 1, l]$ are "critical" for an input A if the overlap remains small in the interval $[g, l]$ *and* if a tape head at time h has moved far from where it was at time g, then it will also move relatively far (and in the same direction) when progressing from time h to time l.

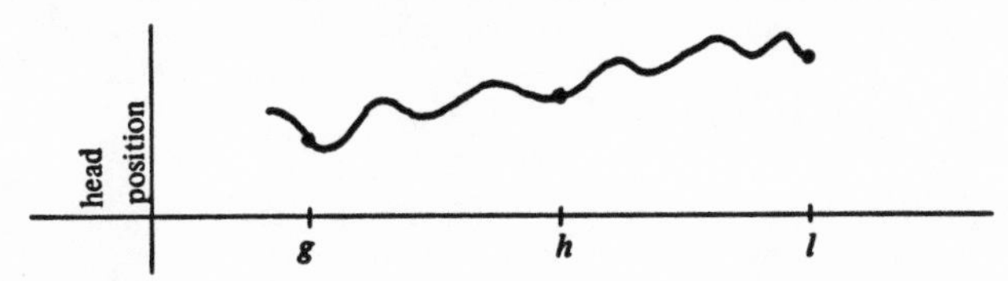

It will follow that if two critical computations are z-equivalent at h, they will be more than z-equivalent at l.

The major steps of the proof are as follows (see §8):

(1) Find an interval I on which the average overlap with respect to S is small. Let $S_0 = \{A \in S|$ overlap for A on I sufficiently exceeds average overlap$\}$ and let $S' = S - S_0$.

(2) Subdivide I into k subintervals $[g_k + 1, g_{k-1}]$, $[g_{k-1} + 1, g_{k-2}]$, $\cdots$, $[g_1 + 1, g_0]$. Define (parameters for) criticalness so that every $A \in S'$ is critical on some $[g_i + 1, g_{i-1}]$ $[g_{i-1} + 1, g_0]$. S_i is the set of all such A. This establishes a (not disjoint) partition of $S' = \bigcup_{1 \leqslant i \leqslant k} S_i$.

(3) Let $E_i^A = \{A_j \in S | A_j$ is the same as A except for possible difference between symbols a_i and b_i on the subinterval $[g_i + 1, g_{i-1}]\}$. Argue that there cannot be too many distinct A_j which are also critical on $[g_i + 1, g_{i-1}]$, $[g_{i-1} + 1, g_0]$. Intuitively, the i.d. equivalence at g_{i-1} must be sufficiently limited else the equivalence at g_0 will be so large as to cause a mistake in distinguishing between $A_1 W \in L_k$ and $A_2 W \notin L_k$ for some appropriate W.

(4) Finally, argue that $\#(E_i^A \cap S_i)$ small implies that $\#S_i$ is small enough to lead to the contradiction that $\Sigma_{i=1}^k \#S_i + \#S_0 < \#S$.

4. Proof of Lemma 1. This section is devoted to the proof of Lemma 1. But we first have to introduce some new definitions, and solve and prove intermediate Lemmas 2–8. Let r be an integer such that $r^{2(r-1)} N_0 \leqslant N < (r + 1)^{2r} N_0$, and let $m = 2(r - 1)$ and let $N_1 = r^m N_0$. Note that since $N > 100 N_0$, we have that $r \geqslant 3$. Given an input interval $I = [g, h]$ let $i^A(I)$ (the dividing point) be the least integer i_0 $(g \leqslant i_0 < h)$ such that $\omega^A(I) = \omega^A([g, i_0], [i_0 + 1, h])$, and also let $\Omega^A(I) = S^A([g, i^A(I)]) \cap S^A([i^A(I) + 1, h])$. Moreover let $\tau^A(I) = \{\langle i, t\rangle | t \in I^{\tau, A}$ and $(\exists s)(\langle i, s\rangle \in \Omega^A(I)$ and t is the first time[2]

[2] Because we are only considering real time machines, it would be sufficient to say $t \in I$ rather than $t \in I^{\tau, A}$.

the square s on tape T_i is scanned in the time interval $([i^A(I) + 1, h])^{\tau,A})\}$.

For each $i = 0, 1, 2, \cdots, m$ we divide the interval $[1, N_1]$ into r^{m-i} successive intervals $I_{i1}, I_{i2}, \cdots, I_{ir^{m-i}}$ of length $N_0 r^i$ each. Thus $I_{ij} = [N_0(j-1)r^i + 1, N_0 jr^i]$; let $i^A_{ij} = i^A(I_{ij})$.

$$\Omega^A_{ij} = \Omega^A(I_{ij}), \quad \omega^A_{ij} = \omega^A(I_{ij}), \quad \tau^A_{ij} = \tau^A(I_{ij}),$$

$$\Xi_i = \{I_{ij} | 1 \leqslant j \leqslant r^{m-i}\},$$

$$\Xi^A_i = \{I_{ij} | I_{ij} \in \Xi_i \text{ and } (\forall h)(\forall s)\ (I_{hs} \in \Xi_h \text{ and } m \geqslant h > i \rightarrow i^A(I_{hs}) \notin I_{ij})\},$$

$$\Xi^{A-}_i = \Xi_i - \Xi^A_i, \quad \Xi = \bigcup_{1 \leqslant i \leqslant m} \Xi_i, \quad \Xi^A = \bigcup_{1 \leqslant i \leqslant m} \Xi^A_i.$$

The following lemma is easy to prove.

LEMMA 2. $\omega^A_{ij} = \#(\tau^A_{ij})$.

PROOF. We shall prove that $\#(\Omega^A(I)) = \#(\tau^A(I))$ which obviously implies Lemma 2. Let f be the following mapping of $\Omega^A(I)$ into $\tau^A(I)$: $f(\langle i, s\rangle) = \langle i, t\rangle$ where $t \in I^{\tau,A}$ and t is the first time the square s on tape T_i is scanned in the time interval $([i^A(I) + 1, N])^{\tau,A}$.

Then f is one-one since a head cannot scan more than one square at each time. Moreover f is also a mapping of $\Omega^A(I)$ onto $\tau^A(I)$ by the definition of $\tau^A(I)$. This implies $\#(\Omega^A(I)) = \#(\tau^A(I))$. Hence, $\omega^A_{ij} = \omega^A(I_{ij}) = \#(\Omega^A(I_{ij})) = \tau^A(I_{ij}) = \tau^A_{ij}$, which proves Lemma 2.

The subset Ξ^A of Ξ was defined with the intent to satisfy the following lemma:

LEMMA 3. *Suppose* $I_{ij} \in \Xi^A$ *and* $I_{uv} \in \Xi^A$, *and suppose* $I_{ij} \neq I_{uv}$. *Then* $\tau^A_{ij} \cap \tau^A_{uv} = \varnothing$.

PROOF. Let $I_{ij} \in \Xi^A$ and $I_{uv} \in \Xi^A$ and suppose $I_{ij} \neq I_{uv}$. Suppose first that $I_{ij} \cap I_{uv} = \varnothing$. Then obviously $\tau^A_{ij} \cap \tau^A_{uv} = \varnothing$. Suppose next that $I_{ij} \cap I_{uv} \neq \varnothing$. Then $I_{ij} \subseteq I_{uv}$ or $I_{uv} \subseteq I_{ij}$. Consider the case $I_{ij} \subseteq I_{uv}$. Then $1 \leqslant i < u \leqslant m$. Moreover by definition of Ξ^A, we have that $i^A(I_{uv}) \notin I_{ij}$. Let $I_{uv} = [g, h]$ and let $i^A(I_{uv}) = l$. Then either $I_{ij} \subseteq [g, l]$ or $I_{ij} \subseteq [l+1, h]$, since $i^A(I_{uv}) \notin I_{ij}$. Suppose first $I_{ij} \subseteq [g, l]$. Then $\langle i, t\rangle \in \tau^A_{ij}$ implies $t \in [g, l]^{\tau,A}$, and $\langle i, t\rangle \in \tau^A_{uv}$ implies $t \in [l+1, h]^{\tau,A}$. Hence $\tau^A_{ij} \cap \tau^A_{uv} = \varnothing$ in this case.

Finally suppose $I_{ij} \subseteq [l+1, h]$, and suppose $\langle i, t\rangle \in \tau_{ij}$. Then there exists a square s on tape T_i such that $\langle i, s\rangle \in S^A(I_{ij})$ and t is the first time s is scanned in the interval $[i_{ij}^A + 1, N]$. But then t is not the first time s is scanned during the interval $[l+1, h]$ since $\langle i, s\rangle \in S^A(I_{ij})$. Hence $\langle i, t\rangle \notin \tau_{uv}$. This proves that $\tau_{ij} \cap \tau_{uv} = \varnothing$ if $I_{ij} \subseteq I_{uv}$. Similarly, $\tau_{ij} \cap \tau_{uv} = \varnothing$ if $I_{uv} \subseteq I_{ij}$. This shows that $\tau_{ij} \cap \tau_{uv} = \varnothing$ in all cases, and Lemma 3 is proved.

The following lemma corresponds to Lemma 11.4, p. 307 in [3]:

LEMMA 4. $N_1 \geqslant (k-1)^{-1}(\Sigma_{I\in\Xi^A}\omega^A(I))$.

PROOF. Let $\tau^A = \bigcup_{I\in\Xi^A}\tau^A(I)$. Then

$$N_1(k-1) \geqslant \#(\tau^A) = \sum_{I\in\Xi^A} \#(\tau^A(I)) \quad \text{(by Lemma 3)}$$

$$= \sum_{I\in\Xi^A} \omega^A(I) \quad \text{(by Lemma 2)}.$$

Hence Lemma 4 follows.

LEMMA 5. $\#(\Xi_i^{A-}) \leqslant (\#(\Xi_i))/(r-1) = r^{m-i}/(r-1)$.

PROOF. Recall that $\Xi_i^{A-} = \Xi_i - \Xi_i^A = \{I_{ij} | I_{ij} \in \Xi_i$ and there exist u and v such that $I_{uv} \in \Xi_u$ and $m \geqslant u > i$ and $i^A(I_{uv}) \in I_{ij}\}$. Note that $\#(\Xi_i) = r^{m-i}$ and $\#(\Xi_u) = r^{m-u}$. The relation $i^A(I_{uv}) \in I$ holds for at most one $I \in \Xi_i$ when $m \geqslant u > i$, and therefore

$$\#(\Xi_i^{A-}) \leqslant \sum_{i<u\leqslant m} \#(\Xi_u) = \sum_{i<u\leqslant m} r^{m-u}$$

$$\leqslant r^{m-i}/(r-1) = \#(\Xi_i)/(r-1).$$

This proves Lemma 5.

LEMMA 6. $mN_1/2(r-1) \geqslant (k-1)^{-1}\Sigma_{I\in\Xi^{A-}}\omega^A(I)$.

PROOF. We have that if $I \in \Xi_i^{A-}$ then $|I| = N_0 r^i$. Hence

$$\omega^A(I) \leqslant |I|(k-1)/2 = N_0 r^i(k-1)/2$$

and

$$\sum_{I\in\Xi^{A-}} \omega^A(I) = \sum_{i=1}^{m} \sum_{I\in\Xi_i^{A-}} \omega^A(I) \leqslant \sum_{i=1}^{m} \#(\Xi_i^{A-}) N_0 r^i (k-1)/2$$

$$\leqslant \sum_{i=1}^{m} \frac{r^{m-i}}{r-1} \frac{N_0 r^i (k-1)}{2} = \frac{(k-1)mN_0 r^m}{2(r-1)} = \frac{(k-1)mN_1}{2(r-1)}.$$

We have proved $(k-1)mN_1/2(r-1) \geqslant \Sigma_{I\in\Xi^A-}\omega^A(I)$ and Lemma 6 is proved.

LEMMA 7. $N_1 \geqslant \frac{1}{2}(k-1)^{-1}\Sigma_{I\in\Xi}\omega^A(I)$.

PROOF.

$$\frac{1}{2(k-1)}\sum_{I\in\Xi}\omega^A(I) = \frac{1}{2(k-1)}\sum_{I\in\Xi^A}\omega^A(I) + \frac{1}{2(k+1)}\sum_{I\in\Xi^A-}\omega^A(I)$$

$$\leqslant \tfrac{1}{2}N_1 + mN_1/4(r-1) \quad \text{(by Lemmas 4 and 6)}$$

$$= \tfrac{1}{2}N_1 + \tfrac{1}{2}N_1 = N_1 \qquad (\text{since } m = 2(r-1)).$$

This proves Lemma 7.

Let $S \subseteq \Gamma^N$; then we have

LEMMA 8. $N_1 \geqslant \frac{1}{2}(k-1)^{-1}\Sigma_{I\in\Xi}\omega^S(I)$.

PROOF. By Lemma 7, we have

$$\#(S)N_1 \geqslant \frac{1}{2(k-1)}\sum_{A\in S}\sum_{I\in\Xi}\omega^A(I)$$

$$= \frac{1}{2(k-1)}\sum_{I\in\Xi}\sum_{A\in S}\omega^A(I)$$

$$= \frac{1}{2(k-1)}\sum_{I\in\Xi}(\#(S))\omega^S(I).$$

Hence we have $N_1 \geqslant \frac{1}{2}(k-1)^{-1}\Sigma_{I\in\Xi}\omega^S(I)$, which proves Lemma 8.

We shall now return to the proof of Lemma 1. Recall that $N_1 = r^{2(r-1)}N_0 \leqslant N < (r+1)^{2r}N_0$ and that $N > 100N_0$ and hence $r \geqslant 3$. Suppose that Lemma 1 were false. Then

$$\xi^S(I) > \frac{4(k-1)\log\log(N/N_0)}{\log(N/N_0)} \qquad \text{for all } I \in \Xi.$$

Hence, we have that

$$\frac{1}{2(k-1)}\sum_{I\in\Xi}\omega^S(I) = \frac{1}{2(k-1)}\sum_{I\in\Xi}\xi^S(I)\cdot|I|$$

$$> \frac{1}{2(k-1)}\sum_{I\in\Xi}\frac{4(k-1)\log\log(N/N_0)\cdot|I|}{\log(N/N_0)}$$

$$= \frac{N_1 m}{2(k-1)}\,\frac{4(k-1)\log\log(N/N_0)}{\log(N/N_0)}$$

$$= 2(N_1 m \log\log (N/N_0)/\log (N/N_0))$$

$$\geqslant N_1 4(r-1) \log\log r^{2(r-1)}/\log (r+1)^{2r}$$

$$= N_1 4(r-1) \log (2(r-1)\log r)/2r \log (r+1)$$

$$= N_1 (2(r-1)/r)(\log (2(r-1)\log r)))/\log (r+1) > N_1$$

(recall $r \geqslant 3$).

Hence, we have proved that $N_1 < \frac{1}{2}(k-1)^{-1}\Sigma_{I \in \Xi}\omega^S(I)$, which contradicts Lemma 8. This proves Lemma 1.

5. The notion of z-equivalence. Let z be a positive integer. We shall say that two i.d.'s

$$D = \langle q, A, t, \alpha, x_1, x_2, \cdots, x_k\rangle \text{ and } D' = \langle q', A', t', \alpha', x'_1, x'_2, \cdots, x'_k\rangle$$

are *z-equivalent* if the following conditions are satisfied.

1. The extended display of D and D' is equal.

2. Suppose $\alpha = \langle\alpha_1, \alpha_2, \cdots, \alpha_k\rangle$ and $\alpha' = \langle\alpha'_1, \alpha'_2, \cdots, \alpha'_k\rangle$. Then we have that for all integers x if $x_j - z \leqslant x \leqslant x_j + z$ then $\alpha_j(x) = \alpha'_j(x), j = 1, 2, \cdots, k$.

Let $A \in \Gamma^N$ and let z, u, v, g, h, l be positive integers such that $0 \leqslant g \leqslant h \leqslant l \leqslant N$. We shall say that the computation of A is $\langle z, u, v, g, h, l\rangle$-*critial* if the following conditions are satisfied.[3]

1. $\omega^A([g, l]) \leqslant v$.

2. If $|\eta_j^A(h+1) - \eta_j^A(g)| > z$ then $|\eta_j^A(l+1) - \eta_j^A(h+1)| > u$ $(j = 1, 2, \cdots, k)$.

LEMMA 9. *Let z, u, v, g, h, l, N be positive integers and let A, A', A_1, X, B, C, Y be words over Γ satisfying the following conditions:*

1. *$v < z$ and $v < u$.*
2. *$0 \leqslant g \leqslant h \leqslant l \leqslant N$.*
3. *$A = A_1XBC$ and $A' = A_1YBC$.*
4. *$g = |A_1| + 1, h = |A_1X| = |A_1Y|$,*
   $l = |A_1XB| = |A_1YB|, N = |A'| = |A|$.
5. *The computation of A and A' are both $\langle z, u, v, g, h, l\rangle$-critical.*
6. *Let $D_0, D_1, \cdots, D_N$ be the computation of A, and let $D'_0, D'_1, \cdots, D'_N$ be the computation of A'. Then D_h and D'_h are $(z+v)$-equivalent.*

Then D_l and D'_l are $(z+u+v)$-equivalent.

---

[3] Note that in case $l = h$, condition 2 will be equivalent to $|\eta_i^A(h+1) - \eta_i^A(g)| \leqslant z$.

PROOF. Note that $D_{g-1} = D'_{g-1}$. Hence $\eta_j^A(g) = \eta_j^{A'}(g)$ for all $j = 1, 2, \cdots, k$. Moreover $\eta_j^A(h+1) = \eta_j^{A'}(h+1)$ for all $j = 1, 2, \cdots, k$, since D_h and D'_h are $(z+v)$-equivalent, which by definition implies that the extended displays of D_h and D'_h are equal. We shall first consider the case that $\eta_j^A(h+1) \geqslant \eta_j^A(g)$.

Then the squares scanned on tape T_i during time interval $[g, h]^{\tau, A}$ is contained in the interval $[\eta_j^A(g) - v, \eta_j^A(h+1) + v]$ since if not then the overlap $\omega^A[g, l]$ will be larger than v violating the fact that the computation of A is $\langle z, u, v, g, h, l\rangle$-critical. Let

$$D_h = \langle q^{(h)}, A, h, \alpha^{(h)}, x_1^{(h)}, x_2^{(h)}, \cdots, x_k^{(h)}\rangle,$$

$$D'_h = \langle p^{(h)}, A', h, \beta^{(h)}, y_1^{(h)}, y_2^{(h)}, \cdots, y_k^{(h)}\rangle.$$

Hence

$$\alpha_j^{(h)}(x) = \beta_j^{(h)}(x) \quad \text{if } x \notin [\eta_j^A(g) - v, \eta_j^A(h+1) + v].$$

Suppose first that $|\eta_j^A(h+1) - \eta_j^A(g)| \leqslant z$. Then since D_h and D'_h are $(z+v)$-equivalent, $\alpha_j^{(h)}(x) = \beta_j^{(h)}(x)$ if $x \in [\eta_j^A(g) - v, \eta_j^A(h+1) + v]$. Hence, in this case we have that $\alpha_j^{(h)}(x) = \beta_j^{(h)}(x)$ for all $x \in Z$. Moreover $q^{(h)} = p^{(h)}$ and $x_i^{(h)} = y_i^{(h)}$ $(i = 1, 2, \cdots, k)$. Then $\alpha_j^{(t)} = \beta_j^{(t)}$ for all $t \geqslant h$ because the same display sequences are generated. To state this claim more precisely, we shall introduce a definition. Consider the sequence $D'_{h+1}, D'_{h+2}, \cdots, D'_N$. Let $e_{h+1}, e_{h+2}, \cdots, e_N$ and $e'_{h+1}, e'_{h+2}, \cdots, e'_N$ be the corresponding extended display sequences. Let (e_n, e'_n) be the first pair such that $(e_n \neq e'_n)$ (i.e., $e_{h+1} = e'_{h+1}, e_{h+2} = e'_{h+2}, \cdots, e_{n-1} = e'_{n-1}$). Since the input is the same after A_1X and A_2Y have been processed, we must have that e_u is of the form $\langle q, a, s_1, s_2, \cdots, s_k, u, x_1, x_2, \cdots, x_k\rangle$ and e_u is of the form $\langle q', a', s'_1, s'_2, \cdots, s'_k, u, y_1, y_2, \cdots, y_k\rangle$ where $q = q', a = a', x_i = y_i$ $(i = 1, 2, \cdots, k)$.

Hence, we must have $s'_i \neq s_i$ for some i. If $s'_i \neq s_i$ we shall say that tape T_i is a d-equivalent stopper in this computation; otherwise T_i is a nonstopper. Going back to the case above we verify immediately that tape T_j cannot be a d-equivalent stopper if $|\eta_j^A(h+1) - \eta_j(g)| \leqslant z$. Suppose next that $\eta_j^A(h+1) \geqslant \eta_j^A(g)$ and $|\eta_j^A(h+1) - \eta_j^A(g)| > z$. Then $|\eta_j^A(l+1) - \eta_j^A(h+1)| > u$, since the computation of A is $\langle z, u, v, g, h, l\rangle$-critical. Moreover, we have that $\eta_j^A(l+1) > \eta_j^A(h+1)$ since otherwise $\omega^A([g, l]) > v$. We also have that the squares scanned during the time interval $[g, h]^{\tau, A}$ and $[g, h]^{\tau, A'}$ are contained in the tape interval $[\eta_j^A(g) - v, \eta_j^A(h+1) + v]$.

Hence $\alpha_j^{(h)}(x) = \beta_j^{(h)}(x)$ for all $x \geqslant \eta_j^A(h+1) - z - v$ since D_h and

D'_h are $(z + v)$-equivalent. Hence, the tape T_j cannot be a d-equivalent stopper in the intervals $[h + 1, l]^{\tau,A}$ and $[h + 1, l]^{\tau,A'}$. The case $\eta_j^A(h + 1) \leq \eta_j^A(g)$ may be dealt with similarly. We have therefore shown that the computation of A and A' are $[h + 1, l]$-equivalent. Hence, $\alpha_j^{(l)} = \beta_j^{(l)}$ if $|\eta_j^A(h + 1) - \eta_j(g)| \leq z$ and $\alpha_j^{(l)}(x) = \beta_j^{(l)}(x)$ for $x \in [\eta_j(l + 1) - z - u - v, \eta_j(l + 1) + z + u + v]$ if $|\eta_j^A(h + 1) - \eta_j(g)| > z$. Hence D_l and D'_l are $(z + u + v)$-equivalent. This proves Lemma 9.

6. The language L_k. We shall now define the language L_k, which can be recognized in real time by a k-pushdown-tape machine but which cannot be recognized in real time by a $(k - 1)$-tape Turing machine. We shall define $L_k = T(M_k)$ where M_k is defined as follows:

$$M_k = \langle \{q_0, q_1\}, \Gamma_k, \{a, b\}, \sigma_k, \lambda_k, b, \{q_0\} \rangle$$

where

$$\Gamma_k^+ = \{a_i | 1 \leq i \leq k\} \cup \{b_i | 1 \leq i \leq k\},$$

$$\Gamma_k^- = \{b'_i | 1 \leq i \leq k\},$$

$$\Gamma_k = \Gamma_k^- \cup \Gamma_k^+.$$

M_k will interpret any of the inputs a_i, b_i and b'_i as some action on tape T_i. M_k will accept the tape depending on the square scanned by the head on tape T_i, and it will write on tape T_i and move the head on tape T_i according to the following rules.

Input a_i *means*: Read the ith working tape. If a is scanned on tape T_i go into state q_1; if b is scanned on tape T_i go into state q_0. Write a on tape T_i and move head number i one square to the right.

b_i *means*: Read the ith working tape. If a is scanned on tape T_i go into state q_1; if b is scanned on tape T_i go into state q_0. Write b on tape T_i and move head number i one square to the right.

b'_i *means*: Read the ith working tape. If a is scanned on tape T_i go into state q_1; if b is scanned on tape T_i go into state q_0. Write b on tape T_i and move head number i one square to the left.

A more formal definition of M_k may be described as follows.

For each $i \in \{1, 2, \cdots, k\}$ let $c \in \{a_i, b_i, b'_i\}$, let $s_j \in \{a, b\}$ $(j = 1, 2, \cdots, k)$ and let $q \in \{q_0, q_1\}$. Let $\epsilon_1 = \epsilon_2 = \cdots = \epsilon_k = 0$. Then

$$\sigma(\langle q, c, s, \cdots, s_{i-1}, a, s_{i+1}, \cdots, s_k \rangle) = q_0,$$

$$\sigma(\langle q, c, s_1, \cdots, s_{i-1}, b, s_{i+1}, \cdots, s_k \rangle) = q_1,$$

$$\lambda(\langle q, a_i, s_1, \cdots, s_{i-1}, s_i, s_{i+1}, \cdots, s_k\rangle)$$
$$= \langle \epsilon_1, \epsilon_2, \cdots, \epsilon_{i-1}, 1, \epsilon_{i+1}, \cdots, \epsilon_k, s_1, \cdots, s_{i-1}, a, s_{i+1}, \cdots, s_k\rangle,$$
$$\lambda(\langle q, b_i, s_1, \cdots, s_{i-1}, s_i, s_{i+1}, \cdots, s_k\rangle)$$
$$= \langle \epsilon_1, \epsilon_2, \cdots, \epsilon_{i-1}, 1, \epsilon_{i+1}, \cdots, \epsilon_k, s_1, \cdots, s_{i-1}, b, s_{i+1}, \cdots, s_k\rangle,$$
$$\lambda(\langle q, b'_i, s_1, \cdots, s_{i-1}, s_i, s_{i+1}, \cdots, s_k\rangle)$$
$$= \langle \epsilon_1, \epsilon_2, \cdots, \epsilon_{i-1}, -1, \epsilon_{i+1}, \cdots, \epsilon_k, s_1, \cdots, s_{i-1}, b, s_{i+1}, \cdots, s_k\rangle.$$

We may interpret the action of M_k as follows: As long as symbols Γ_k^+ are entering, store them on the tape as prescribed. When the symbols in Γ_k^- appear, move back and retrieve the stored symbols, while entering accepting or nonaccepting states depending on the stored information.

As can easily be seen by inspection M_k is a k-pushdown tape machine in our sense, since M_k always writes b (= blank) when moving left.

REMARKS ON THE CONJECTURE. To prove the conjecture $C_k - C'_{2k-1} \neq 0$, we would propose to use a language obtained by adding $\{a'_i \mid i = 1, 2, \cdots, k\}$ to the input alphabet. a'_i would then mean the same action as a_i except that we shall move the head on tape T_i to the left instead of to the right.

7. The class S of input words. Recall

$$\Gamma_k^+ = \{a_i \mid 1 \leqslant i \leqslant k\} \cup \{b_i \mid 1 \leqslant i \leqslant k\},$$
$$\Gamma_k^- = \{b'_i \mid 1 \leqslant i \leqslant k\},$$
$$\Gamma_k = \Gamma_k^- \cup \Gamma_k^+.$$

We shall think of k as fixed throughout the rest of the paper. Hence, we shall write Γ, Γ^- and Γ^+ instead of Γ_k, Γ_k^- and Γ_k^+, respectively. Let $A \in (\Gamma^+)^*$. Then $(A)_i$ is the word obtained from A by deleting every a_j and b_j in A if $j \neq i$. (*Example.* $(a_1a_2b_1a_3b_2)_2 = a_2b_2$.) B is a *subword* of A if there exist words A_1 and A_2 (possibly empty) such that $A_1BA_2 = A$. We shall now define a class of input words $S \subseteq (\Gamma^+)^N$ (for some positive integer N), such that the frequencies of the letters a_1 and b_1 is ρ^{i-1} times the frequencies of the letters a_i and b_i $(i = 2, 3, \cdots, k)$, where ρ is a positive integer > 2. We shall do so by first defining a master word $W_0 \in \{a_i \mid 1 \leqslant i \leqslant k\}^N$, and then defining S to be the class of all words obtained from W_0 by replacing some occurrences of a_i by b_i $(i = 1, 2, \cdots, k)$. Given $\rho > 2$, let $\zeta = \rho^{-1}$ and let $\Theta = 1 + \rho + \rho^2 + \cdots + \rho^{k-1} = (\rho^k - 1)/(\rho - 1)$; $\theta = 1 + \zeta + \zeta^2 + \cdots + \zeta^{k-1} = (1 - \zeta^k)/(1 - \zeta) = \Theta\rho^{1-k}$. Then let W_1 be a word such that

$W_1 \in \{a_i | 1 \leqslant i \leqslant k\}^{\Theta}$ and $|(W_1)_i| = \rho^{k-i}$ $(i = 1, 2, 3, \cdots, k)$. (By the definition of Θ we can easily verify that such a word exists.) Let W_0 be the segment of W_1^N of length N. Hence, $|W_0| = N$ and $W_0 = W_1 W_1 \cdots W_1 W_1'$, where W_1' is an initial segment of W_1. We now define S as follows: $S = \{W | W$ is obtained from W_0 by replacing some occurrences of a_i by b_i $(i = 1, 2, \cdots, k)\}$. The following lemmas follow immediately from the definition.

LEMMA 10. $\#(S) = 2^N$.

PROOF. Let $W \in S$. Then $|W| = N$ and let $W_0 = c_1 c_2 \cdots c_N$ and $W = c_1' c_2' \cdots c_N'$, where $c_j, c_j' \in \Gamma^+$ $(j = 1, 2, \cdots, N)$. Then, for each c_j $(j = 1, 2, \cdots, N)$, we have that $c_j = a_i$ for some i. Hence, we have for c_j' a choice of two letters a_i and b_i. Hence, the total number of such $W \in S$ is 2^N. This proves Lemma 10.

LEMMA 11. *Let $A \in S$ and let B be a subword of A such that $|B|/\Theta$ is an integer. Then we have, for every $i = 1, 2, 3, \cdots, k$,*

$$|(B)_i| = \theta^{-1}\rho^{1-i}|B|.$$

PROOF. By the construction of S, it is sufficient to prove that Lemma 11 holds if $A = W_0$. Hence, assume $A = W_0$. Then there exist words B_1 and B_2 such that $W_1 = B_1 B_2$ and $B = B_2 W_1 W_1 \cdots W_1 B_1$, since $|B|/|W_1|$ is an integer. But, by construction, $|(W_1)_i| = \rho^{k-i}$ and $|W_1| = \Theta = \theta\rho^{k-1}$. Hence $|(W_1)_i| = \theta^{-1}\rho^{1-i}|W_1|$. Hence $|(B)_i| = \theta^{-1}\rho^{1-i}|B|$. This proves Lemma 11.

8. The completion of the proof. From now on we shall assume that we have a $(k-1)$-tape real time Turing machine $M = (K, \Gamma, \Sigma, \sigma, \lambda, q_0, b, F)$ where $\Gamma = \Gamma_k$, which recognizes L_k. We shall prove that this leads to a contradiction. Let[4]

$$\rho = 8k(k-1)(\log_2(\#\Sigma) + 1), \qquad \zeta = 1/\rho, \tag{3}$$

$$\Theta = 1 + \rho + \rho^2 + \cdots + \rho^{k-1} = (\rho^k - 1)/(\rho - 1). \tag{4}$$

$$\theta = 1 + \zeta + \zeta^2 + \cdots + \zeta^{k-1} = (1 - \zeta^k)/(1 - \zeta) = \Theta\rho^{1-k}. \tag{5}$$

(Then $\Theta = \theta\rho^{k-1}$.)

$$N_0 = 32k(k-1)^2\theta^2\rho^{2k}([\log_2\#(K)] + 1)^2. \tag{6}$$

---

[4] Assume without loss of generality that $\#\Sigma = 2^{n_1}$ and $\#K = 2^{n_2}$ for some n_1 and n_2.

(Then $N_0/\Theta k$ is an integer.) Choose N such that

$$(7) \qquad (4(k-1)\log\log(N/N_0))/\log(N/N_0) \leqslant 4^{-1}\rho^{-k-2}k^{-1}$$

(choosing $N = q^q N_0$, where $q = 32(k-1)k\rho^{k+2}$ will suffice). Choose the input interval $I = [g+1, l]$ such that $|I|/N_0$ is an integer and $0 \leqslant g \leqslant h \leqslant N$ and such that $\xi^S(I) \leqslant (4(k-1)\log\log(N/N_0))/\log(N/N_0)$. Then, by (7),

$$(8) \qquad \xi^S(I) \leqslant 4^{-1}\rho^{-k-2}k^{-1}.$$

Since $|I|$ is divisible by N_0 which is again divisible by $k\Theta = k\theta\rho^{k-1}$, we can write I as $[g+1, g+hk]$, where h again is divisible by Θ. Then $|I| = hk$. Let

$$(9) \qquad g_i = g + h(k-i) \qquad (i = 0, 1, 2, \cdots, k),$$

$$(10) \qquad I_i = [g_i + 1, g_{i-1}] \qquad (i = 1, 2, \cdots, k).$$

Note that $g_k = g$, and that $g_0 = g + hk = l$, and I is the same as $[g_k + 1, g_0]$. Moreover, since $h = |I_i|$ we have that $|I_i|/\Theta$ is an integer. Let

(11) $v = \rho^{-k-2}h$ (note that since kh/N_0 is an integer, v is an integer),

(12) $z_i = h\theta^{-1}\rho^{-i}k$ $(i = 1, 2, 3, \cdots, k)$,

(13) $u_i = z_{i-1} - 2v$ $(i = 2, 3, \cdots, k)$,

(14) $u_1 = h\theta^{-1} - 2v$.

By (8) we have that

$$(15) \qquad \omega^S(I) \leqslant v/4$$

since

$$\omega^S(I) \leqslant 4^{-1}\rho^{-k-2}k^{-1}|I| = 4^{-1}\rho^{-k-2}k^{-1}kh = v/4.$$

Let

$S_0 = \{A | A \in S \text{ and } \omega^A(I) \geqslant v\}$,

$S' = S - S_0 = \{A | A \in S \text{ and } \omega^A(I) < v\}$,

$S_i = \{A | A \in S'$ and the computation of A is $\langle z_i, u_i, v, g_i + 1, g_{i-1}, g_0\rangle$-critical$\}$;

that is, since $A \in S'$ implies $\omega^A(I) < v$ which means $\omega^A([g_i + 1, g_0]) < v$, we have $S_i = \{A | A \in S'$ and if $|\eta_j(g_i + 1) - \eta_j^A(g_{i-1} + 1)| > z_i$ then $|\eta_j^A(g_0 + 1) - \eta_j^A(g_{i-1} + 1)| > u_i\}$.

LEMMA 12. $\#(S_0) \leqslant \#(S)/4 = 2^{N-2}$.

The proof follows immediately from (15) and the definition of S_0, and Lemma 10.

LEMMA 13. $S = S_0 \cup S_1 \cup \cdots \cup S_k$.

PROOF. Suppose $A \in S$ but $A \notin (S_0 \cup S_1 \cup \cdots \cup S_k)$. Then $A \in S$, and $\omega^A(I) < v$. For each $i = 1, 2, 3, \cdots, k$ let $j(i)$ be the least j such that $|\eta_j(g_{i-1}+1) - \eta_j(g_i+1)| > z_i$ and $|\eta_j^A(g_0+1) - \eta_j^A(g_{i-1}+1)| \leqslant u_i$. (Such a j exists since $A \notin S_i$.) We shall prove that $j(i_1) \neq j(i_2)$ if $i_1 \neq i_2$. Suppose $i < r$ and $j = j(i) = j(r)$. Then $g_r < g_{r-1} \leqslant g_i < g_{i-1} \leqslant g_0$. Then we have that

$$|\eta_j^A(g_0+1) - \eta_j^A(g_{r-1}+1)| \leqslant u_r = z_{r-1} - 2v.$$

We also have that if $g_{r-1} + 1 \leqslant x \leqslant y \leqslant g_0 + 1$ then

$$|\eta_j^A(y) - \eta_j^A(x)| < u_r + 2v = z_{r-1} \leqslant z_i,$$

for if not then $\omega^A(I) \geqslant v$. Hence, $|\eta_j^A(g_{i-1}+1) - \eta_j(g_i+1)| < z_i$ which contradicts the fact that $j = j(i)$. Hence, we have proved that if $i < r$ then $j(i) \neq j(r)$, which implies that $j(i_1) \neq j(i_2)$ if $i_1 \neq i_2$. Consider the set $P = \{j(1), j(2), \cdots, j(k)\}$. Since $j(i_1) \neq j(i_2)$ if $i_1 \neq i_2$ we must have that $\#(P) = k$; but $P \subseteq \{1, 2, \cdots, k-1\}$. Hence $\#(P) \leqslant k-1$, which is a contradiction. (This is the only place in the proof where we use the fact that M has fewer than k tapes in an essential way.) This completes the proof of Lemma 13.

Let $A \in S$, let $0 \leqslant g \leqslant h \leqslant N = |A|$ and let $I = [g+1, h]$. Then we shall use the expressions $A([g+1, h]) = A(I)$ to denote the substring C_2 of A where C_1, C_2 and C_3 are chosen such that $A = C_1C_2C_3$ and $|C_1| = g$ and $|C_1C_2| = h$.

Let $A_1 \in S$ and $A_2 \in S$. Then we define an equivalence relation $\equiv_i$ for each $i = 1, 2, \cdots, k$ as follows.

$A_1 \equiv_i A_2$ holds if the following two conditions are satisfied:

1. $A_1([1, g_i]) = A_2([1, g_i]), A_1([g_{i-1}+1, N]) = A_2([g_{i-1}+1, N])$.

2. Let B_1 and B_2 be the result of replacing every b_i by a_i in $A_1(I_i)$ and $A_2(I_i)$, respectively. Then $B_1 = B_2$.

Given $A \in S$, let

$$E_i^A = \{B | B \in S \text{ and } B \equiv_i A\} \qquad (i = 1, 2, \cdots, k),$$

$$F_i = \{B | B \in S \text{ and } (B(I_i))_i \in \{a_i\}^*\} \quad (i = 1, 3, \cdots, k).$$

(Here $\{a_i\}^* = \{e, a_i, a_ia_i, a_ia_ia_i, \cdots\}$ where e is the empty word.)

The following four lemmas follow immediately from the definitions.

LEMMA 14. $\#(E_i^A \cap F_i) = 1$ *for all* $i = 1, 2, \cdots, k$ *and all* $A \in S$.

LEMMA 15. $\equiv_i$ *is an equivalence relation* $(i = 1, 2, \cdots, k)$.

LEMMA 16. $S = \bigcup_{A \in F_i} E_i^A \quad (i = 1, 2, 3, \cdots, k)$.

LEMMA 17. $S_i = \bigcup_{A \in F_i} (E_i^A \cap S_i) \quad (i = 1, 2, 3, \cdots, k)$.

Let $d(i) = \theta^{-1}\rho^{1-i}h$; then we have

LEMMA 18. *Let* $A \in S$. *Then*

$$|(A(I_i))_i| = \theta^{-1}\rho^{1-i}h = d(i)$$

and

$$|A([g_i + 1, g_0])_i| = i\theta^{-1}\rho^{1-i}h = id(i).$$

PROOF. Since $|I_i| = h$ and Θ divides h we have by Lemma 11 that

$$|(A(I_i))_i| = \theta^{-1}\rho^{1-i}|A(I_i)| = \theta^{-1}\rho^{1-i}h = d(i).$$

In the same way we get $|(A[g_i + 1, g_0])_i| = i\theta^{-1}\rho^{1-i}h = id(i)$. This proves Lemma 18.

LEMMA 19. $\#(E_i^A) = 2^{d(i)} \quad (i = 1, 2, \cdots, k)$ *and* $\#(F_i) = 2^{N-d(i)}$ $(i = 1, 2, \cdots, k)$.

PROOF. If $A_1 \in E_i^A$ and $A_2 \in E_i^A$ then, according to the definition of E_i^A and Lemma 18, A_1 and A_2 can differ at most at $d(i)$ places each containing either a_i or b_i. Hence $\#(E_i^A) = 2^{d(i)}$.

In the same way, if $A_1 \in F_i$ and $A_2 \in F_i$ then A_1 and A_2 can differ at most at $N - d(i)$ places. Hence $\#(F_i) = 2^{N-d(i)}$. This proves Lemma 19.

LEMMA 20. *Let* $e(i) = 2(k-1)(z_i + v)$ *and let* $A \in S$. *Then* $\#(E_i^A \cap S_i) \leqslant (\#(\Sigma))^{e(i)}(\#(K)) \cdot (2h)^{k-1}$.

PROOF. Let $A \in S$. Then we shall use $D(A, t)$ here to be the $t + 1$ instantaneous description of the computation of A. Hence, $D(A, 0)$ is the first i.d. and $D(A, N)$ is the last i.d. in the computation of A when $|A| = N$. We shall first use Lemma 9 to prove that $A_1 \in E_i^A \cap S_i$ and $A_2 \in E_i^A \cap S_i$, and if $A_1 \neq A_2$ then $D(A_1, g_{i-1})$ and $D(A_2, g_{i-1})$ are not $(z_i + v)$-equivalent. Suppose the contrary. Suppose $A_1 \in E_i^A \cap S_i$ and $A_2 \in E_i^A \cap S_i$; suppose $A_1 \neq A_2$; and suppose $D(A_1, g_{i-1})$ and $D(A_2, g_{i-1})$ are $(z_i + v)$-equivalent.

Then, by Lemma 9, $D(A_1, g_0)$ and $D(A_2, g_0)$ are $(z_i + u_i + v)$-equivalent. We have that

$$|(A_1([g_i + 1, g_0]))_i| = |A_2([g_i + 1, g_0])_i| = i\theta^{-1}\rho^{1-i}h = id(i),$$

by Lemma 18. Hence, since $A_1 \neq A_2$ and $A_1 \equiv_i A_2$, we must have that $A_1([1, g_0])(b_i')^x \in L_k$ iff $A_2([1, g_0])(b_i')^x \notin L_k$, for some nonnegative integer $x \leqslant id(i)$. But $z_i = h\theta^{-1}\rho^{-i}k$, $v = \rho^{-k-2}h$ and $u_i = h\theta^{-1}\rho^{1-i}k - 2v$. Hence, $x \leqslant i\theta^{-1}\rho^{1-i}h \leqslant h\theta^{-1}\rho^{1-i}k = z_{i-1} = u_i + 2v < u_i + z_i + v$ since $z_i > v$. Hence, $x < u_i + z_i + v$. But since $D(A_1, g_0)$ and $D(A_2, g_0)$ are $(z_i + u_i + v)$-equivalent, we must have that $A_1([1, g_0])(b_i')^x \in T(M)$ iff $A_2([1, g_0])(b_i')^x \in T(M)$. This contradicts the fact that $T(M) = L_k$. Hence, we have proved that if $A_1 \in E_i^A \cap S_1, A_2 \in E_1^A \cap S_i$ and $A_1 \neq A_2$, then $D(A_1, g_{i-1})$ and $D(A_2, g_{i-1})$ are not $(z_i + v)$-equivalent. Let $\mathcal{D}_i = \{D(A_1, g_{i-1}) | A_1 \in E_i^A \cap S_i\}$. Then $\#(\mathcal{D}_i) = \#(E_i^A \cap S_i)$, and $\mathcal{D}_i$ is a set of non-$(z_i + v)$-equivalent i.d.'s. Since $D(A_1, g_i) = D(A_2, g_i)$ if $A_1 \in E_i^A \cap S_i$ and $A_2 \in E_i^A \cap S_i$ we have that the number of different head positions is at most $(2h)^{k-1}$ since each head cannot move more than h squares during interval I_i from the position the head had in the i.d. $D(A_1, g_i)$. Moreover, the number of different states is $\#(K)$. Finally for each head position the number of different tape expressions which are nonequal in the tape interval $[\eta_j^A(g_{i-1} + 1) - z_i - v, \eta_j^A(g_{i-1} + 1) + z_i + v]$ is $(\#(\Sigma))^{e(i)}$ where $e(i) = 2(k-1)(z_i + v)$. Multiplying all these numbers we get that $\#(\mathcal{D}_i) \leqslant (\#(\Sigma))^{e(i)} \#(K) \cdot (2h)^{k-1}$. Since $\#(\mathcal{D}_i) = \#(E_i^A \cap S_i)$, Lemma 20 is proved.

Let $\mu_i^A = \#(E_i^A \cap S_i)/\#(E_i^A)$. Then we have

LEMMA 21. *$\mu_i^A < 1/(2k)$ for $i = 1, 2, \cdots, k$ and all $A \in S$.*

PROOF. We shall prove that $\log_2(\mu_i^A) < -\log_2(2k)$, which will prove Lemma 21.

By Lemmas 19 and 20 we have that

$$\log_2(\mu_i^A) \leqslant e(i) \log_2(\#(\Sigma)) + \log_2(\#(K)) + (k-1)\log_2(2h) - d(i).$$

Here $d(i) = \theta^{-1}\rho^{1-i}h$ and $e(i) = 2(k-1)(z_i + v)$. Hence we have that

$$2\log_2(\mu_i^A) \leqslant 4(k-1)z_i \log_2(\#(\Sigma)) - d(i) + R \tag{16}$$

where

$$R = (4v(k-1))\log_2(\#(\Sigma)) + 2\log_2(\#(K)) + 2(k-1)\log_2(2h) - d(i). \tag{17}$$

We shall show that $R < 0$. We shall first prove that

$$4v(k-1)\log_2(\#(\Sigma)) < d(i)/4. \tag{18}$$

We have that

$$4v(k-1)\log_2(\#(\Sigma)) < 4\rho^{-k-2}h(k-1)([\log_2(\#(\Sigma))]+1)$$
$$= \rho^{-k-1}h/2k < \theta^{-1}\rho^{1-i}h/4 = d(i)/4 \quad \text{(by (3))}.$$

Hence (18) is proved. Next we are going to prove that

$$(19) \qquad 2\log_2(\#(K)) < d(i)/4.$$

We have that N_0 divides $k \cdot h$; hence

$$d(i) = \theta^{-1}\rho^{1-i}h \geqslant 32(k-1)^2\theta^2\rho^{2k}([\log_2 \#(K)]+1)^2\theta^{-1}\rho^{1-i}$$
$$= 32(k-1)^2\theta\rho^{2k+1-i}([\log_2 \#(K)]+1)^2 > 8\log_2(\#(K)).$$

Hence (19) follows. Finally we shall prove

$$(20) \qquad 2(k-1)\log_2(2h) < d(i)/2 = (\theta^{-1}\rho^{1-i}h)/2.$$

Hence we have to prove that

$$(21) \qquad 2h/\log_2(2h) > 8(k-1)\theta\rho^{i-1}.$$

But $2h = 8^2(k-1)^2\theta^2\rho^{2k}y^2$ for some real number $y > 1$. Let $x = (2h)^{1/2}$. Then

$$2h/\log_2(2h) = x^2/\log_2(x^2) = x(x/2\log_2 x) > x \quad \text{since } x > 4$$

and therefore $x/2\log_2 x > 1$. Hence

$$2h/\log_2(2h) > x = 8(k-1)\theta\rho^k y > 8(k-1)\theta\rho^{i-1}$$

and (20) follows.

From (18), (19) and (20) it follows that $R < 0$. Hence by (16) we have

$$(22) \qquad 2\log_2(\mu_i^A) < 4(k-1)z_i\log_2(\#(\Sigma)) - d(i).$$

We shall next prove that

$$(23) \qquad 8(k-1)z_i\log_2(\#(\Sigma)) < d(i).$$

We have that

$$8(k-1)z_i\log_2(\#(\Sigma)) < 8(k-1)h\theta^{-1}\rho^{-i}k([\log_2(\#(\Sigma))]+1) = h\theta^{-1}\rho^{1-i} = d(i).$$

Hence (23) follows. By (22) and (23) we have

$$\log_2(\mu_i^A) < -d(i)/4 = -\theta^{-1}\rho^{1-i}h/4 \leqslant -\theta^{-1}\rho^{1-i}N_0/(4k)$$
$$< -2(k-1)^2\theta\rho^{2k+1-i} < -\log_2(2k).$$

Hence Lemma 21 is proved.

LEMMA 22. $\#(S_i) < 2^N/(2k)$.

PROOF.

$$\begin{aligned}
\#(S_i) &\leq \sum_{A \in F_i} \#(E_i^A \cap S_i) && \text{(by Lemma 17)} \\
&= \sum_{A \in F_i} \mu_i^A \#(E_i^A) && \text{(by definition)} \\
&< \sum_{A \in F_i} \frac{1}{2k} \#(E_i^A) && \text{(by Lemma 21)} \\
&= \sum_{A \in F_i} \frac{2^{d(i)}}{2k} && \text{(by Lemma 19)} \\
&= 2^{N-d(i)} \frac{2^{d(i)}}{2k} && \text{(by Lemma 19)} \\
&= 2^N/(2k).
\end{aligned}$$

This proves Lemma 22.

LEMMA 23. $\#(S_1 \cup S_2 \cup \cdots \cup S_k) < 2^{N-1}$.

PROOF. Lemma 23 follows immediately from Lemma 22.

But we now have a contradiction. By Lemma 13 we have that $S = S_0 \cup S_1 \cup \cdots \cup S_k$, and by Lemmas 12 and 23 we have that $\#(S) \leq 3 \cdot 2^{N-2}$. But $\#(S) = 2^N$ by Lemma 10. This proves that $T(M) \neq L_k$. Hence no $(k-1)$-tape real time Turing machine can recognize L_k, which by definition can be recognized by a k-pushdown-tape real time machine. This proves Theorem 1.

Acknowledgment. I am very grateful to Professor Michael Paterson for helpful discussion. I would also like to thank Professor Allan Borodin for help in preparing this paper, and for help in proofreading the typed copy of this paper.

References

1. A. Cobham, *The intrinsic computational difficulty of functions*, Proc. 1964 Internat. Congress for Logic, Methodology, and Philosophy of Science, North-Holland, Amsterdam, 1965, pp. 24–30. MR 34 #7376.

2. S. A. Cook, *On the minimum computation time of functions*, Doctoral Thesis, Harvard University, Cambridge, Mass., 1966.

3. S. A. Cook and S. O. Aanderaa, *On the minimum computation time of functions*, Trans. Amer. Math. Soc. 142 (1969), 291–314. MR 40 #2459.

4. P. C. Fischer, A. R. Meyer and A. L. Rosenberg, *Time-restricted sequence generation*, J. Comput. System Sci. 4 (1970), 50–73. MR 40 #6808.

5. S. Ginsburg, *The mathematical theory of context-free languages*, McGraw-Hill, New York, 1966. MR 35 #2692.

6. J. E. Hopcroft and J. D. Ullman, *Formal languages and their relation to automata*, Addison-Wesley, Reading, Mass., 1969. MR 38 #5533.

7. J. Hartmanis and R. E. Stearns, *On the computational complexity of algorithms*, Trans. Amer. Math. Soc. 117 (1965), 285–306. MR 30 #1040.

8. F. C. Hennie, *One-tape, off-line Turing machine computations*, Information and Control 8 (1965), 553–578. MR 32 #9171.

9. F. C. Hennie and R. E. Stearns, *Two-tape simulation of multitape Turing machines*, J. Assoc. Comput. Mach. 13 (1966), 533–546. MR 35 #1413.

10. M. Paterson, M. Fischer and A. R. Meyer, *An improved overlap argument for on-line multiplication*, SIAM-AMS Proc., vol. 7, Amer. Math. Soc., Providence, R. I., 1974. pp. 97–125.

11. M. O. Rabin, *Real-time computation*, Israel J. Math. 1 (1963), 203–211. MR 29 #1148.

12. A. L. Rosenberg, *Real-time definable languages*, J. Assoc. Comput. Mach. 14 (1967), 645–662. MR 38 #3099.

UNIVERSITY OF OSLO

SIAM–AMS Proceedings
Volume 7
1974

An Improved Overlap Argument For On-Line Multiplication*

Michael S. Paterson, Michael J. Fischer and Albert R. Meyer

Abstract. A lower bound of $cN \log N$ is proved for the mean time complexity of an on-line multi-tape Turing machine performing the multiplication of N-digit binary integers. For a more general class of machines which includes some models of random-access machines, the corresponding bound is $cN \log N/\log \log N$. These bounds compare favorably with known upper bounds of the form $cN(\log N)^k$, and for some classes the upper and lower bounds coincide. The proofs are based on the "overlap" argument due to Cook and Aanderaa.

1. Introduction. A challenging problem in the field of computational complexity is to prove lower bounds on the computing time for naturally defined algorithms executed by realistically powerful machinery. For a serial machine whose task is to map an input string to an output string, a trivial lower bound for many mappings is the number of steps required to read the input string. There are a number of combinatorial techniques, involving for example crossing sequences (cf. [6, §10.4]), which are adequate to derive nontrivial lower bounds but only for rudimentary machines such as single-tape Turing machines. The powerful diagonalization techniques are of use only for an input/output mapping sufficiently structured to encode machine computations.

In this paper we expound and develop further the "overlap" argument

---

AMS (MOS) subject classifications (1970). Primary 68A20.

*Much of this work was carried out at the University of Warwick with partial support from the Science Research Council of the United Kingdom. It was also supported in part by the National Science Foundation under research grant GJ-34671 to MIT Project MAC, and in part by the Artificial Intelligence Laboratory, an MIT research program sponsored by the Advanced Research Projects Agency, Department of Defense, under Office of Naval Research contract number N00014-70-A-0362-0003.

introduced by Cook and Aanderaa [3], which establishes a nonlinear lower bound on the time required by a very general class of machines to perform multiplication of binary integers. (A similar argument has been used again recently by Aanderaa [1].) Our contribution relative to [3] is firstly that the main line of proof is somewhat shortened and simplified, secondly that the lower bound in [3] is increased by a factor $\log \log n$ and is shown to hold for the average rather than just the worst case, and thirdly a new observation is made which yields an even stronger result for the case of multi-tape Turing machines. In some cases we show our new results to be optimal to within a constant factor by exhibiting suitable multiplication algorithms.

Our main results are lower bounds for "on-line" multiplication. A mapping from an input string to an output string is said to be carried out *on-line* if for all n the nth output symbol is printed after the nth and before the $(n + 1)$st input symbol is read. [1] For on-line multiplication the multipliers and multiplicands are given in binary, least significant digit first, and each input symbol encodes the two corresponding input digits. We may as well assume that, for N-digit arguments, only the least significant N digits of the product are to be produced. (The remaining digits may be obtained, if desired, by concatenating N zeros to the arguments.) On-line multiplication is of course possible, though a naive implementation may take time at least proportional to n between the $(n - 1)$st and nth digits, with therefore a time of order N^2 for an N-digit product. We show here that the minimum average computation time for on-line multiplication is bounded below and above by functions of the form $N(\log N)^k$, where, for the lower bounds, k is approximately 1 and, for the upper bounds, k is approximately 1 or 2 depending on the class considered. The exact results are given in §§6, 7 and 8. For further background and motivation the reader is referred to [3].

2. Machine models. In the class of machines to which our proofs apply, we wish to include not only the familiar multi-tape Turing machine but also Turing machines with tapes of higher dimension and some suitably tame "random access machines." We shall have to exclude iterative arrays and other machines with

[1] For technical convenience, we use this strong form of the definition which prohibits an output from being produced too soon. It is not a serious restriction for two reasons. For binary multiplication, the ith digit of the product cannot be determined until the ith digits of the two inputs have been read except when both numbers are even; hence a machine can take advantage of the weaker definition for at most a quarter of all possible inputs, changing our mean time bounds by only a constant factor. Secondly any BAM may be modified without time loss to obey the strong definition by adding a two-headed linear tape to serve as an output buffer. This does not affect any of our lower bounds. (Cf. [5].)

unlimited parallelism since these are able to do multiplication in "real-time" [2]. Our definitions follow [3] fairly closely, with minor differences in order to simplify the notation and proof or to take fuller advantage of the power of the proof technique. The reader is assumed to have experience with the basic definitions and techniques of automata theory [6].

A *bounded activity machine* (BAM) has a deterministic finite-state control which operates with a one-way read-only input tape, a one-way write-only output tape and a *storage structure.* The storage structure is a countable set of *locations* each of which can hold a binary value. The store is accessed and modified by a finite, fixed number of *work heads* whose moves are specified by a finite set of *shifts* $\varphi_1, \cdots, \varphi_p$. For each i, φ_i is a map from the set of locations into itself, and a head at some location x may be moved in one step to the location $\varphi_i(x)$. A complete step of the BAM is described as follows. Depending on the state of the finite control, the input tape *may* be advanced one symbol and precisely one work head *must* be "moved" by one of the shifts. Then, depending on the control state, the new input symbol (if any), and the value stored in the storage location to which the head is moved, a new value may be stored, an output symbol may be given, and a new control state is entered. Thus, for each given symbol from the input tape, there is a unique step at which it is read, and the definition prevents it from being reread later. Moreover, only one storage symbol is read per step, so we may speak of "the storage symbol at step s."

Various restricitions in this definition, such as binary storage, one head move per step, and the lack of dependence of the new step on the old storage values, are introduced to simplify the exposition and cause a time penalty of at most a constant factor compared with more versatile machines.

We shall say that a computation is *real-time* if it is on-line and each input symbol is read a fixed number of steps after the previous input. Note that we shall not require the store to be initially "empty" except for the special class of "uniform" machines defined below.

It is easy to design a BAM which can multiply in real-time. A suitable storage structure is based on an infinite binary tree, traversed by a single head which takes left or right branches depending on the input digits. The correct output is either already stored in the location of the tree at the start or else is encoded in the structure itself in an obvious way. For example the structure may have a shift ψ such that, for all x, either $\psi(x) = x$ or else $\psi(\psi(x)) = x$ and $\psi(x) \neq x$. Which alternative holds can be determined for any location by a sequence of a few steps.

Two classes introduced by Cook and Aanderaa [3] to evade such an oracular construction are the *polynomial-limited* and *uniform* machines defined below. We also add two further classes.

(i) *Polynomial-limited.* A storage structure is *polynomial-limited* if there are constants c, d such that, for all locations x and for all t, the number of locations accessible from x in t steps is no greater than ct^d. A BAM with such a structure is a *polynomial-limited machine.*

(ii) *Uniform.* A storage structure is *uniform* if, for each pair of locations x, y, there is a permutation f such that $f(x) = y$ and, for each shift φ_i of the structure, $f \circ \varphi_i = \varphi_i \circ f$. A BAM with a uniform structure which is initially "empty" (i.e., each location has the same initial value) is a *uniform machine.* The reader is referred to [3] for further discussion of these and other classes.

With a suitable form of definition, Turing machines, even with multiple heads and multi-dimensional tapes, satisfy *both* restrictions. Our main result holds for machines satisfying *either* restriction. The BAM described above which multiplies in real-time satisfies neither.

(iii) *One-dimensional multi-head multi-tape Turing machines.* We can obtain a stronger result than for classes (i) and (ii) if we restrict the tapes to be linear, i.e., one-dimensional.

(iv) *Oblivious machines.* For this class we turn our attention from the storage structure, which may be arbitrary, to the form of the finite state control or "program." The (single) storage location accessed at each step defines the *storage sequence* for any computation, and this depends in general on the input. A machine is *oblivious* if for input sequences of a given length the storage sequence is fixed, i.e., independent of the input symbols. Naturally, the control state and the values inscribed in the store can, and in general do, depend on the input symbols; it is just the movement of heads which is invariant. Our interest in oblivious machines is two-fold. Firstly the restriction permits a very simple proof of an improved lower bound, and secondly it happens that almost all the algorithms proposed or used for multiplication are oblivious or can be made oblivious at the cost of only a constant factor in time.

In §7 we shall retrospectively consider other classes of machines to which the proof technique applies.

3. Retrorse functions. Informally, a function from an input string to the output is *retrorse* [2] if the output values in any segment depend very heavily on the input values of the immediately preceding segment, and so the function evaluator needs to "turn back" to the previous input segment. On-line multiplication will be shown to be very retrorse.

[2] We choose not to follow Cook and Aanderaa in their choice of "complex" to describe these functions.

We define $K_N = \Sigma_{2^i<N} 2^{2^i}$, so the binary expansion of K_N has a "1" in position i iff i is a power of two, where the positions are numbered starting with 0 at the right (lower-order) end. The usefulness of K_N is that multiplication of N-digit numbers by K_N is extremely retrorse, and the main proof is simpler than for two-input multiplication. Our first theorem provides a lower bound on the average time for on-line multiplication of an N-digit integer by K_N and hence also on the worst-case time for on-line multiplication. In the second theorem we show that the same bound holds for the average time for general on-line multiplication.

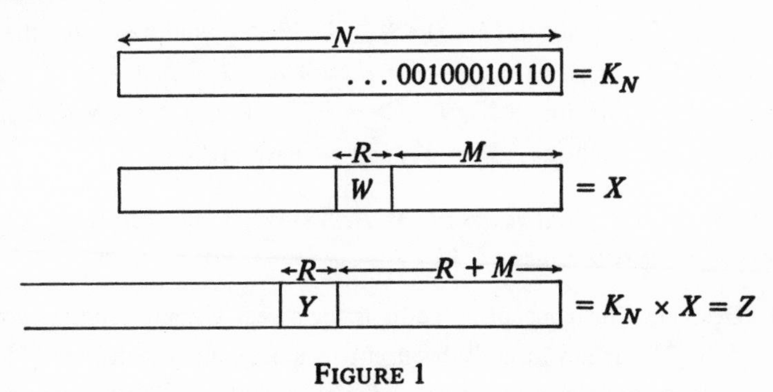

FIGURE 1

Figure 1 represents the multiplication of K_N by an N-digit number X with result Z. It is drawn in the conventional way with least significance to the right. R and M are nonnegative integers, and we shall always take R to be a power of two, 2^r. W represents the subfield of X consisting of bits $X_{M+R-1} \cdots X_{M+1}X_M$, and Y represents the subfield of Z consisting of bits $Z_{M+2R-1} \cdots Z_{M+R+1}Z_{M+R}$. We will at times think of W and Y as R-bit integers in the range 0 to $2^R - 1$. To say that W assumes a value i means that we imagine placing the binary representation of i into the W subfield of X. This in turn causes Z to change, for Z always means the product $K_N \cdot X$, and that in turn affects the value of Y. We investigate the dependence in this way of Y upon W.

The way in which Y varies with W of course depends on the remaining bits of X, other than those in W, which we denote by $X \backslash W$. As a number, $X \backslash W$ is the value of the binary string obtained by setting the W-field of X to zero.

For some particular fixed value of $X \backslash W$, let W range through all possible values $0, 1, \cdots, 2^R - 1$, so $X_i = X\backslash W + i \cdot 2^M, 0 \leqslant i \leqslant 2^R - 1$. Let $Z_0, Z_1, \cdots$ and $Y_0, Y_1, \cdots$ be the corresponding values of Z and Y, that is, $Z_i = K_N \cdot X_i$, and Y_i is the Y-field of Z_i.

If $i < j$, then $Z_j - Z_i = (X_j - X_i) \cdot K_N = (j - i) \cdot 2^M \cdot K_N$. Since $K_N = K_{2R} + 2^{2R} \cdot \bar{K}$ for some integer $\bar{K}$, we have

$$Z_j - Z_i \equiv (j - i) \cdot 2^M \cdot K_{2R} \pmod{2^{M+2R}}.$$

Now suppose $Y_i = Y_j$. Then $Z_j - Z_i \equiv a \cdot 2^M \pmod{2^{M+2R}}$ for some integer a, $|a| < 2^R$, and hence $(j - i) \cdot K_{2R} \equiv a \pmod{2^{2R}}$. Since $R = 2^r$,

$$2(2^R - 1) \geqslant K_{2R} = 2^{2^r} + 2^{2^{r-1}} + \cdots + 2^{2^0} \geqslant 2^R.$$

By the right-hand inequality, $(j - i) \cdot K_{2R} \geqslant 2^R > a$, and hence for the congruence to hold, we must have $(j - i) \cdot K_{2R} \geqslant a + 2^{2R} > 2^{2R} - 2^R$. From the left-hand inequality for K_{2R}, $j - i > \frac{1}{2} \cdot 2^R$. Hence, for any i, there is at most one $j > i$ such that $Y_i = Y_j$, and we have proved

LEMMA 1. *For fixed values of* M, R, N *and* $X \backslash W$, *each value of* Y *can arise from at most two values of* W.

4. Overlap. This concept is the basis for a very elegant counting argument introduced in [3]. It has recently been put to use again by Aanderaa [1]. The motivation for its definition comes from the computation of very retrorse functions. The obvious way in which information about a previous input segment can be obtained is by revisiting locations which were visited when that segment was being read. Overlap is defined in terms only of the storage sequence defined previously in §2(iv). If two successive accesses to the same location l occur at steps s_1 and s_2 $(s_1 \leqslant s_2)$, then the pair (s_1, s_2) is called an *overlap pair*, l is called the *overlap location* of s_2, and the value stored in l at step s_1 and referenced at step s_2 is called the *overlap value* of s_2. The *total overlap* is

$$\Omega = |\{(s_1, s_2) | (s_1, s_2) \text{ is an overlap pair}\}|.$$

Clearly the total time $T \geqslant \Omega$, since each step s is the second component of one or zero overlap pairs depending on whether the location accessed at step s has been accessed before or not.

Let C_1, C_2 be disjoint contiguous time intervals during a computation. We define *overlap* (C_1, C_2) to be the number of overlap pairs (s_1, s_2) for which $s_1 \in C_1$ and $s_2 \in C_2$.

Without loss of results we assume $N = 2^n$. For any $i = 0, \cdots, n$, which we call the *level*, define $R_i = 2^i$, and if $S = S_{N-1}S_{N-2} \cdots S_1S_0$ is any string of length N, we partition S into contiguous blocks $S_{i,2^{n-i}-1}, \cdots, S_{i,1}, S_{i,0}$ of length R_i, where $S_{i,j} = S_{(j+1)R_i - 1} \cdots S_{j \cdot R_i}$, $0 \leqslant j \leqslant 2^{n-i} - 1$.

If X is the input string, the time interval $C_{i,j}$ starts as the right-most digit of $X_{i,j}$ is read and continues until the right-most digit of $X_{i,j+1}$ is about to be read (or until the computation ends if there is no such $j+1$).

Let $t_{i,j}$ be the length of time $C_{i,j}$. Clearly the total time of the computation $T = \Sigma_j t_{i,j}$ for each level i. We define $\omega_{i,j} =$ overlap $(C_{i,j}, C_{i,j+1})$ for all suitable i, j, and also $\omega_i = \Sigma_j \omega_{i,2j}$ for any i.

LEMMA 2. *Total overlap* $\Omega = \Sigma_i \omega_i$.

PROOF. Let (s_1, s_2) be an overlap pair and let i be the least level such that $s_1, s_2 \in C_{i,j}$ for some j. $C_{i,j}$ is the concatenation of the two intervals $C_{i-1,2j}$ and $C_{i-1,2j+1}$ at the next lower level. By our choice of i, $s_1 \in C_{i-1,2j}$ and $s_2 \in C_{i-1,2j+1}$, so (s_1, s_2) contributes to $\omega_{i-1,2j}$ and hence to ω_{i-1}. Suppose it contribues to $\omega_{i',2j'}$. Then $i' \leqslant i-1$, for s_1 and s_2 belong to the same interval for each level above $i-1$. $i' \geqslant i-1$ since if (s_1, s_2) contributes to $\omega_{i',2j'}$, then both s_1 and s_2 are in the same block $C_{i'+1,j'}$ at level $i'+1$. Hence $i' = i-1$, and it is clear that $j' = j$. We conclude that each overlap pair contributes exactly once to exactly one ω_i and hence exactly once to $\Sigma_i \omega_i$. $\square$

5. Computations with small overlap. We consider an on-line computation of some machine $\mathfrak{M}$ from input X to output Z. As before, M and R are fixed numbers, W is the length R subword of X, namely $X_{M+R-1} \cdots X_M$, and Y is the length R subword of Z, namely $Z_{M+2R-1} \cdots Z_{M+R}$. Throughout this section, $X \backslash W$, M and R remain fixed, and we explore how the value of Y changes as the value of W is varied. Unlike the previous sections, Z now represents the output of $\mathfrak{M}$ on input X.

Giving a particular value to W completely determines the computation of $\mathfrak{M}$. Let s_0 be the step which advances the input head onto the first symbol of W, let s_1 be the step which moves the input head off of the last symbol of W, and let s_2 be the step which reads the next input symbol after producing Y. Define interval C_W to be the steps from s_0 to $s_1 - 1$, C_Y the steps from s_1 to $s_2 - 1$, and let $t_W = s_1 - s_0$ and $t_Y = s_2 - s_1$ be the lengths of time associated with C_W and C_Y, respectively. T is the total time of the computation, and we let $\omega =$ overlap(C_W, C_Y). This notation is illustrated in Figure 2.

As W varies, so do ω, t_Y, t_W, T and Y. Let $Q(\hat{\omega}, \hat{t}, \hat{T})$ be the total number of different Y values yielded by those W such that $\omega \leqslant \hat{\omega}$, $t_Y \leqslant \hat{t}$, and $T \leqslant \hat{T}$. If $\mathfrak{M}$ is computing a retrorse function, $Q(\hat{\omega}, \hat{t}, \hat{T})$ must be large, and we will use this fact to deduce the constraints on $\hat{\omega}$, $\hat{t}$, and $\hat{T}$ that eventually lead to our lower bound on T.

Our upper bounds on Q depend on the kind of machine, but all are obtained using the same general method. For a given value of W, we observe the computation during the interval C_Y and we record in a suitable way information about W that affects the computation in C_Y.

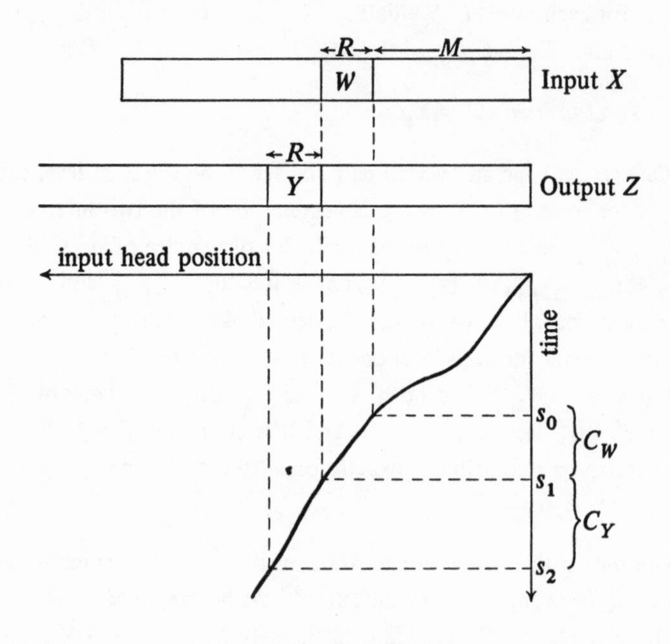

FIGURE 2

For each class of machines, enough information will be recorded to ensure the validity of the following:

CONDITION 1. Let w and w' be two values for W and y and y' be the corresponding values of Y. If the information recorded for w and w' is the same, then $y = y'$.

It follows immediately from Condition 1 that the number of different possible information records obtained from values of W for which ω, t_Y, and T are bounded respectively by $\hat{\omega}$, $\hat{t}$, and $\hat{T}$ is an upper bound on $Q(\hat{\omega}, \hat{t}, \hat{T})$.

LEMMA 3. *There exists a constant* C *depending only on* $\mathfrak{M}$ *such that, for* $\hat{T} > 1$ *and* $\hat{\omega} \leq \hat{t} \leq \hat{T}$,

(a) $Q(\hat{\omega}, \hat{t}, \hat{T}) \leq \hat{T}^C \cdot 2^{\hat{\omega}} \cdot \binom{\hat{t}}{\hat{\omega}}$ *if* $\mathfrak{M}$ *is a polynomial-limited or uniform*;

(b) $Q(\hat{\omega}, \hat{t}, \hat{T}) \leq \hat{T}^C \cdot 2^{\hat{\omega}}$ *if* $\mathfrak{M}$ *is a one-dimensional multi-head multi-tape Turing machine*;

(c) $Q(\hat{\omega}, \hat{t}, \hat{T}) \leq C \cdot 2^{\hat{\omega}}$ *if* $\mathfrak{M}$ *is oblivious.*

PROOF. (a) *Case* 1. $\mathfrak{M}$ *is a polynomial-limited*. The information record consists of the state of the control, the position of each head at time s_1, a subset $\theta = \{t_1, \cdots, t_{\hat{\omega}}\}$ of the integers from 0 through $\hat{t} - 1$, and a binary sequence v of length $\hat{\omega}$. θ is chosen to include the times, relative to s_1, of all the W-overlap steps. The ith bit of v equals the symbol referenced at time $s_1 + t_i$, which will be the overlap value if $s_1 + t_i$ is a W-overlap step. This insures that Condition 1 is satisfied.

The total number of such records is clearly at most $c \cdot 2^{\hat{\omega}} \cdot \binom{\hat{t}}{\hat{\omega}} \cdot H^c$, where H is the number of possible positions for each head at step s_1. Since $\mathfrak{M}$ is polynomial-limited, $H \leqslant t_W^d \leqslant T^d \leqslant \hat{T}^d$, yielding the bound of part (a).

(a) *Case* 2. $\mathfrak{M}$ *is uniform.* This case is exactly like Case 1 except that we do not record the actual head position at time s_1, for the number of possible positions is too large. Rather, we record for each head h the step of C_Y (if any) at which h first visits a square l visited prior to step s_0 together with the time of the first visit to l (which uniquely specifies l). Call such a location l *filled.* (In the case of more than one head, a small amount of additional information must be recorded to account for the possible interactions among the heads before revisiting a filled location. This argument is presented in more detail in [3].)

If the head h never visits a filled location during C_Y, then because of uniformity the symbols read and written by the head can be uniquely determined from the overlap steps and values, without knowing the position of h at time s_1. On the other hand, if h does visit a filled location l, then the step of C_Y at which l is visited together with l itself provide all the information required to determine the symbols read by the head during C_Y. l can be specified by the time of its first visit, so there are at most $t_Y \cdot T \leqslant \hat{T}^2$ different starting positions of a single head h which must be distinguished.

(b) $\mathfrak{M}$ *is a one-dimensional multi-head multi-tape Turing machine.* $\mathfrak{M}$ is a special case of a polynomial-limited machine, so we may record the state and starting head positions as in (a), Case 1. However, the positions at which overlap will occur may be specified much more succinctly, for the squares visited during C_W by each head form an interval. Thus, only the endpoints need be named. Since at most $2T$ locations on a linear tape can be reached in T steps by a given head, there are at most $2T^2$ possible intervals per head. As before, a binary sequence of length $\hat{\omega}$ is sufficient in which to record the overlap values. Thus, the total number of such records is at most $c \cdot H^c \cdot (2\hat{T}^2)^c \cdot 2^{\hat{\omega}}$, where H is as in (a), Case 1.

(c) $\mathfrak{M}$ *is oblivious.* The positions of the heads at each step are independent of W, so only the state of the control at step s_1 and the values of the overlap locations need to be recorded, giving the bound $C \cdot 2^{\hat{\omega}}$. □

We finish our preparations for the main proof with a combinatorial lemma.

By way of motivation, let $\mathfrak{M}$ be a machine that multiplies on-line and let $\hat{\omega}$, $\hat{t}$, and $\hat{T}$ be bounds on ω, t_Y, and T respectively such that $Q(\hat{\omega}, \hat{t}, \hat{T}) < 2^{3R/4}$. By Lemma 1, all but $2 \cdot 2^{3R/4}$ values of W, a vanishing fraction of the 2^R values, cause one of the three bounds to be exceeded. This gives an implicit lower bound on ω in terms of t_Y and T which says in effect that if T is small, then the total overlap Ω is large, which implies that T is large. Hence, T must be large on the average.

LEMMA 4. *Let C, R, a be positive constants such that* $\log a > 2(C + 3)$. *If* $0 < \omega \leqslant t$ *and* $\omega + t/a + \log T \leqslant R/(2 \log a)$, *then* $T^C \cdot 2^\omega \cdot \binom{t}{\omega} < 2^{3R/4}$. *(All logarithms in this paper are taken to base 2.)*

PROOF. For any $p \geqslant q > 0$,

$$\binom{p}{q} \leqslant \frac{p^q}{q!} \leqslant \left(\frac{pe}{q}\right)^q$$

by Stirling's formula. Assume the hypothesis, so $\omega < R/(2 \log a)$, $t < aR/(2 \log a)$, and $\log T < R/(2 \log a)$.

$$T^C \cdot 2^\omega \cdot \binom{t}{\omega} < T^C \cdot \binom{2t}{\omega}$$

which is monotonic increasing in ω, t and T since $\omega \leqslant t$. So

$$T^C \cdot \binom{2t}{\omega} < 2^{CR/(2 \log a)} \cdot \binom{aR/\log a}{R/(2 \log a)}$$

$$< 2^{CR/(2 \log a)} \cdot (2ae)^{R/(2 \log a)} < 2^{3R/4}. \quad \square$$

6. Main results and proofs.

THEOREM 1. *There is a constant* c *such that for any BAM* $\mathfrak{M}$ *which, for all N multiplies N-digit numbers by* K_N *on-line, the mean time* $T(N)$ *over all numbers of length* N *satisfies the following bounds for all sufficiently large* N.

(i) *If* $\mathfrak{M}$ *is polynomial-limited or uniform,* $T(N) > cN \log N/\log \log N$.

(ii) *If* $\mathfrak{M}$ *is a one-dimensional multi-head and multi-tape Turing machine or is oblivious,* $T(N) > cN \log N$.

PROOF. Suppose first that $\mathfrak{M}$ is polynomial limited or uniform. We may assume that $\log \log N > 2(C + 3)$ where C is as in Lemma 3, and define $a = \log N$. We consider again the situation depicted in Figures 1 and 2, where $X \backslash W$ is fixed and W is allowed to vary. Applying Lemmas 1, 3(a) and 4, we deduce that the number of distinct W's for which $\omega + t/a + \log T \leqslant R/(2 \log a)$ is less than $2 \cdot 2^{3R/4}$. Hence certainly,

$$\text{mean}_W\ (\omega + t/a + \log T) > R/(3 \log a) \quad \text{for } R \geqslant 16.$$

Since this inequality holds for all values of $X \backslash W$,

$$\text{mean}_X\ (\omega + t/a + \log T) > R/(3 \log a)$$

where the mean is taken over all N-digit numbers. If we assume that $\text{mean}_X\ T < N^2$ then $\text{mean}_X \log T \leqslant \log \text{mean}_X\ T < 2 \log N$ since the geometric mean is less than or equal to the arithmetic mean. Now use the identity: $\text{mean}(A + B) = \text{mean}(A) + \text{mean}(B)$. Therefore

$$\text{mean}_X\ (\omega + t/a) > R/(3 \log a) - 2 \log N > R/(4 \log a)$$

provided that $R > 24 \cdot (\log N) \cdot (\log \log N)$.

Now we suppose $W = X_{i,2j}$ and $Y = Z_{i,2j+1}$ for some i, j, so that $\omega = \omega_{i,2j}$, $t = t_{i,2j+1}$ and $R = R_i = 2^i$. Taking the intervals in pairs by summing over j, the previous inequality gives

$$\begin{aligned}\text{mean}_X\ \omega_i + \text{mean}_X \sum_j t_{i,2j+1}/a &= \sum_j \text{mean}_X\ (\omega_{i,2j} + t_{i,2j+1}/a) \\ &> \sum_j R_i/(4 \log a) = N/(8 \log a).\end{aligned}$$

If we assume that $\text{mean}_X\ T < N \log N/(16 \log \log N)$, then since $\Sigma_j\, t_{i,2j+1} < T$, we have

$$\begin{aligned}\text{mean}_X\ \omega_i &> N/(8 \log a) - N/(16 \log \log N) \\ &= N/(16 \log \log N).\end{aligned}$$

Since this inequality holds for all i such that $i < \log N$ and $2^i = R_i > 24 \cdot (\log N) \cdot (\log \log N)$, we conclude that

$$\begin{aligned}\text{mean}_X\ T > \text{mean}_X\ \Omega &= \sum_i \text{mean}_X\ \omega_i \\ &\geqslant [\log N - \log(24(\log N)(\log \log N))] \cdot N/(16 \log \log N) \\ &\geqslant N \cdot \log N/(17 \log \log N)\end{aligned}$$

provided N is sufficiently large. Thus, we have proved case (i).

The proof for case (ii) is somewhat simpler. We can easily show that if $\omega + 2C \cdot \log T \leqslant R/2$ then $T^C \cdot 2^\omega < 2^{3R/4}$. From this we deduce in a similar way to case (i) that $\text{mean}_X\ (\omega + 2C \cdot \log T) > R/3$. If $\text{mean}_X\ T < N^2$ and $R_i > 48 \cdot C \cdot \log N$ then $\text{mean}_X\ \omega_{\ i,2j} > R_i/4$ and $\text{mean}_X\ \omega_i > N/8$, so

$$\text{mean}_X\ T > \text{mean}_X\ \Omega = \sum_i \text{mean}_X\ \omega_i > N \cdot (\log N)/9.$$

This proves case (ii). Of course a proof solely for oblivious machines would be very easy since ω_{ij}, t_{ij}, T, etc. are independent of X. □

We can immediately extend this proof, removing the dependence on K_N. We show that nearly all numbers as multipliers yield a function nearly as retrorse as multiplication by K_N. In the situation of Figure 1 and Lemma 1, let us replace K_N by an arbitrary N-digit number K_N^*.

LEMMA 5. *For any* h, $0 < h < 2^R$, *if, for some* i, $Y_i = Y_{i+h}$, *then* K_{2R}^* *must have one of at most* 2^{R+1} *possible values.*

PROOF. As in the proof of Lemma 1, if $Y_i = Y_{i+h}$, then $h \cdot K_{2R}^* \equiv a (\text{mod } 2^{2R})$ for some a, $|a| < 2^R$.

Let $d = \gcd(h, 2^{2R})$. Then $d|a$, so $a = kd$, where $|kd| < 2^R$. Also, $d|2^{R-1}$ by definition of d and the fact that $h < 2^R$. Hence,

$$k \in \left\{-\frac{2^R}{d} + 1, -\frac{2^R}{d} + 2, \cdots, -1, 0, 1, \cdots, \frac{2^R}{d} - 1\right\}$$

so there are $(2 \cdot 2^R/d) - 1$ such k's.

By elementary number theory, there are exactly d values of K_{2R}^* in the range $0 \leq K_{2R}^* < 2^{2R}$ which satisfy $hK_{2R}^* \equiv kd (\text{mod } 2^{2R})$. Hence, there are at most $d \cdot (2 \cdot 2^R/d - 1) < 2^{R+1}$ values of K_{2R}^* for which $Y_i = Y_{i+h}$. □

From this lemma, it follows at once that at least half of all possible values of K_{2R}^* have the property that, for all h, $0 < h \leq 2^{R-2}$, and for all i, $Y_i \neq Y_{i+h}$. Hence, for these values of K_{2R}^*, at most 4 different W's yield the same Y. Therefore the proof of Theorem 1 can be followed very closely except that "mean_X" is replaced throughout by "$\text{mean}_{K_N^*} \text{mean}_X$". Thus we have shown

THEOREM 2. *There is a constant* c *such that, for any BAM* $\mathfrak{M}$ *which performs on-line multiplication, the mean time* $T(N)$ *for pairs of N-digit numbers satisfies the following bounds for all sufficiently large* N.

(i) *If* $\mathfrak{M}$ *is polynomial-limited or uniform,*

$$T(N) > c N \log N/\log \log N,$$

(ii) *If* $\mathfrak{M}$ *is a multi-tape Turing machine or is oblivious,*

$$T(N) > c N \log N.$$

We know of no direct implication between Theorems 1 and 2.

7. Extensions. In this section we shall outline some of the ways in which the classes already considered can be extended while remaining susceptible to the same proof methods.

A simple "random-access" machine could be modelled by a BAM with a storage structure based upon some sort of binary tree so that locations are "addressed" by binary sequences and accessed in time proportional to the address length. Such a structure is of course exponentially, rather than polynomially, limited. However we recall that the latter property is used in the proof only to allow head positions to be specified just before a new input symbol is to be read. The proof goes through just as before therefore, provided that the tree structure is used in such a way that the heads are returned to the root before each input is read, for example, if all the "random-accessing" is accomplished by a subroutine.

An alternative approach to a random-access store is the structure based on the free group on two generators a, b with the four shifts being left multiplication by a, a^{-1}, b, b^{-1}. This can be operated as a quite serviceable random-access store and is of course uniform.

A point of merely technical interest is that the same bounds may be easily proved when the polynomial limited class is extended by replacing "ct^d" in the definition by "$c2^{t^\epsilon}$" for any $\epsilon < 1$. Unfortunately we know of no natural class of machines which takes advantage of this extension.

Finally we show that without impairing the proof of any of the four classes of machines we may add "oracles," and indeed more, in the following way. The BAM's are extended by allowing an infinite number of states in the control subject only to the restriction that just a finite number of them may read the input tape. An "oracle," which may even be nonrecursive, could be invoked with such a machine to read a sequence of storage locations and put the result of applying its oracular function in some other sequence of locations. This would take just the number of steps required to access the locations. The proofs are unaffected by this relaxation. A simple example which emphasizes the importance to our proof of the on-line restriction is a multi-tape Turing machine with an oracle to perform (off-line) multiplication in linear time. The $cN \log N$ lower bound applies even to this machine.

8. Upper bounds. An important technique for establishing upper bounds for on-line multiplication is given by M. Fischer and Stockmeyer [4]. Their construction shows that, for a wide range of machine classes including multi-tape Turing machines, oracle Turing machines, and oblivious machines, given any off-line (i.e., unrestricted) multiplication machine with time complexity $T(N)$, where T satisfies $T(2N) \geqslant 2T(N)$, an on-line machine can be produced with time complexity no greater than $c \cdot T(N) \cdot \log N$.

A slight extension of their methods shows that on-line multiplication of N-digit integers, where one of the numbers has at most $\log N$ "1"-digits, can be performed in time $O(N \cdot \log N)$. In particular, there is a Turing machine for on-line multiplication by K_N with complexity $O(N \cdot \log N)$, matching the bound of Theorem 1 (ii).

General on-line multiplication algorithms may be obtained by applying the Fischer-Stockmeyer result to the off-line algorithm of Schönhage-Strassen [7], which on a Turing machine has complexity $O(N \cdot \log N \cdot \log \log N)$. With the facility of constructing and rapidly accessing a multiplication table for $(\log N)$-digit numbers such as is provided by a random-access machine, the Schönhage-Strassen off-line multiplication algorithm can be performed in time $O(N \cdot \log N)$.

	Multi-Tape TM	Oracle TM	"RAM"
On-line multiplication	$N \log^2 N \log \log N$	$\underline{N \log N}$	$N \log^2 N$
On-line multiplication by K_N	$\underline{N \log N}$	$\underline{N \log N}$	$N \log N$
Off-line multiplication	$N \log N \log \log N$	$\underline{N}$	$N \log N$

FIGURE 3

In Figure 3 we set out some of the upper bounds derived from the above results for three classes of machines. Constant factors are omitted and underlining denotes that a lower bound of the same order has been demonstrated in previous sections. The first class is multi-tape Turing machines with one-dimensional tapes; the second is BAM's with an infinite number of states under the restriction on input states given in §7; the third class is either version of "random-access" machine described in §7. With the uniform structure based on the free group on two generators, it is easy to simulate Turing machines and stores with "random-access." BAM's with the binary tree structure and the restriction on head positions given in §7 are also sufficiently powerful to allow a fast implementation of the required algorithms, though the programming techniques needed are less straightforward.

9. Conclusion. In this paper we have described a powerful counting argument based on the notion of "overlap" and have investigated the extent and limitations of its applicability. Overlap arguments are applicable only under the on-line restriction, but in many cases they can lead to complexity bounds which are optimal within a constant factor.

An important objective for future research is to obtain nontrivial lower bounds without the severe restriction to on-line computation. Such results, even for oblivious machines or combinatorial circuits, would constitute a significant advance.

References

1. S. O. Aanderaa, *On k-tape versus (k + 1)-tape real-time computation*, SIAM-AMS Proc., vol. 7, Amer. Math. Soc., Providence, R. I., 1974, pp. 75–96.

2. A. J. Atrubin, *A one-dimensional real-time iterative multiplier*, IEEE Trans. Electronic Computers EC–14 (1965), 394–399.

3. S. A. Cook and S. O. Aanderaa, *On the minimum computation time of functions*, Trans. Amer. Math. Soc. 142 (1969), 291–314. MR 40 #2459.

4. M. J. Fischer and L. J. Stockmeyer, *Fast on-line integer multiplication*, Proc. 5th ACM Sympos. on Theory of Computing, 1973, pp. 67–72; J. Comput. Sys. Sci. 9 (1974) (to appear).

5. P. C. Fischer, A. R. Meyer and A. L. Rosenberg, *Real-time simulation of multihead tape units*, J. ACM 19 (1972), 590–607.

6. J. E. Hopcroft and J. D. Ullman, *Formal languages and their relation to automata*, Addison-Wesley, Reading, Mass., 1969. MR 38 #5533.

7. A. Schönhage and V. Strassen, *Schnelle Multiplikation grosser Zahlen*, Computing (Arch. Elektron. Rechnen) 7 (1971), 281–292. MR 45 #1431.

UNIVERSITY OF WARWICK

MASSACHUSETTS INSTITUTE OF TECHNOLOGY

SIAM-AMS Proceedings
Volume 7
1974

String-Matching and Other Products*

Michael J. Fischer and Michael S. Paterson

Abstract. The string-matching problem considered here is to find all occurrences of a given pattern as a substring of another longer string. When the pattern is simply a given string of symbols, there is an algorithm due to Morris, Knuth and Pratt which has a running time proportional to the total length of the pattern and long string together. This time may be achieved even on a Turing machine. The more difficult case where either string may have "don't care" symbols which are deemed to match with all symbols is also considered. By exploiting the formal similarity of string-matching with integer multiplication, a new algorithm has been obtained with a running time which is only slightly worse than linear.

1. Introduction. We consider several problems concerned with the matching of strings of symbols. A typical practical problem is that we are given a (long) symbol string $\mathbf{X} = X_0X_1X_2 \cdots X_m$, the "text," and another (short) string $\mathbf{Y} = Y_0Y_1 \cdots Y_n$, the "pattern," over the same finite alphabet Σ. The task is to find all occurrences of the pattern as a consecutive substring in the text, that is, to find all i, $n \leqslant i \leqslant m$, such that

$$\mathbf{Y} = [X_{i-n} \cdots X_i].$$

The obvious naive algorithm tries each i in turn and compares Y_j with X_{i-n+j} for $j = 0, 1, \cdots$ as far as necessary, and is represented by the following informal program:

AMS (MOS) subject classifications (1970). Primary 68A20, 68A10.

*Much of this work was carried out at the University of Warwick with partial support from the Science Research Council of the United Kingdom. It was also supported in part by the National Science Foundation under research grant GJ-34671 to MIT Project MAC, and in part by the Artificial Intelligence Laboratory, an MIT research program sponsored by the Advanced Research Projects Agency, Department of Defense, under Office of Naval Research contract number N00014-70-A-0362-0003.

```
FOR i = n STEP 1 UNTIL m
    FOR j = 0 STEP 1 UNTIL n
        IF Y_j ≠ X_{i-n+j} GOTO L
    REPEAT
    PRINT (i)
L: REPEAT
```

For example with the following strings the desired outputs would be 4, 7, 12.

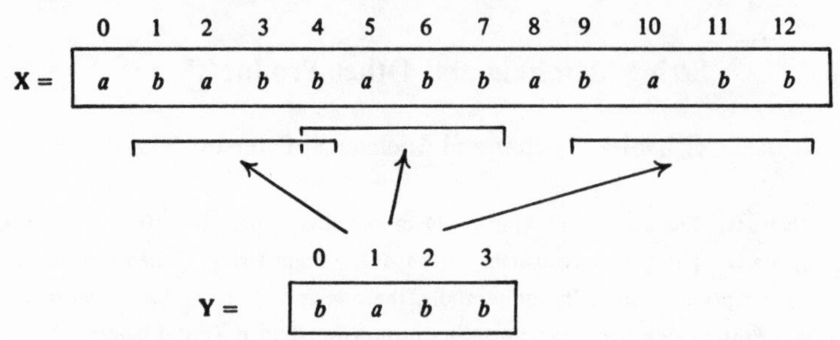

An upper bound on the computation time for this algorithm is $O(m \cdot n)$ and the matching of $a^m b$ with $a^n b$ shows that this bound is realistic.

2. Morris-Knuth-Pratt algorithm. A considerable improvement on the naive procedure described above is afforded by an algorithm due to J. H. Morris, D. E. Knuth and V. R. Pratt [4], which has a running time which is $O(m + n)$. The essential idea is that if we have successfully matched a segment of the string **X** with an initial segment of **Y** before reaching an inequality, then it is unnecessary and wasteful to read those symbols of **X** again since they are *the same as the* **Y**-*segment*. A better procedure is to carry out the first comparisons for the next relative position of the pattern **Y**, by comparing **Y** with a segment of *itself*, and of course the comparisons can be precomputed once and for all at the beginning. The precomputation required is very quick and has the same general form as the main computation itself.

We shall describe a "theoreticians' version" of the Morris-Knuth-Pratt algorithm to simplify the presentation and analysis. For a symbol string $\mathbf{Z} = Z_0 \cdots Z_n$, define the function P for $i = 0, \cdots, n$ by

$$P(i) = \max \left\{ t \,\middle|\, \bigwedge_{0 \leqslant r \leqslant t} Z_{i-r} = Z_{t-r} \text{ and } -1 \leqslant t < i \right\}.$$

Provided we consider $\bigwedge_{0 \leqslant r \leqslant -1}$ to be identically *true*, $P(i)$ is always well defined. It is not difficult to verify that $P^{(k)}(i) = k$th largest t such that $\bigwedge_{0 \leqslant r \leqslant t} Z_{i-r} = Z_{t-r}$ and $-1 \leqslant t < i$ if this is defined. ($P^{(k)}(i)$ denotes the composition of P with itself k times, so $P^{(0)}(i) = i$ and $P^{(k+1)}(i) = P(P^{(k)}(i))$.) The usefulness of P results from the following recursive definition.

$$P(0) = -1,$$

$$P(i+1) = \begin{cases} P^{(k)}(i) + 1 & \text{for } 0 \leqslant i < n \text{ where } k \text{ is the least positive integer} \\ & \text{such that } Z_{[P^{(k)}(i)+1]} = Z_{i+1}, \\ -1 & \text{if there is no such } k. \end{cases}$$

An example is illustrated below.

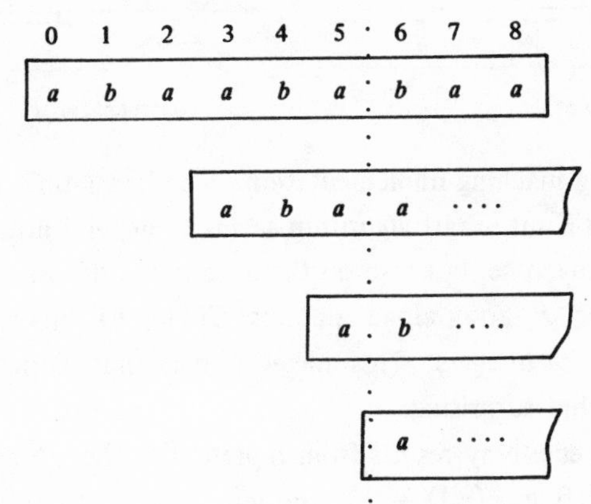

$P(5) = 2, \quad P(P(5)) = 0,$

$P^{(3)}(5) = -1, \quad P(6) = P(P(5)) + 1 = 1.$

i	0	1	2	3	4	5	6	7	8
$P(i)$	-1	-1	0	0	1	2	1	2	3

Provided that the string **Z** and the values of $P(j)$ for $j \leqslant i$ are readily acces-sible, the value of $P(i + 1)$ may be computed in time less than

$$c \cdot (P(i) - P(i + 1) + 2)$$

for some constant c, independent of i and n. This is because

$$P(j + 1) - 1 \leqslant P(j) < j \quad \text{for all } j$$

and hence the k of the recursive definition satisfies

$$P(i) - P(i + 1) + 2 \geqslant k \geqslant 1.$$

Therefore the total running time is bounded by

$$c \cdot (P(0) - P(n)) + 2c(n + 1) = O(n).$$

To solve our original problem we concatenate **Y**, a new symbol @, and **X** in that order and compute the values of P for the string **Y @ X** in time $O(m + n)$. Because of the @, P can never take a value greater than $n = |\mathbf{Y}| - 1$. The values of i for which $P(i) = n$ mark the positions where **Y** matches a substring of **X**, or more precisely

$$P(n+2+i) = n \Leftrightarrow \bigwedge_{0 \leqslant r \leqslant n} X_{i-r} = Y_{n-r}.$$

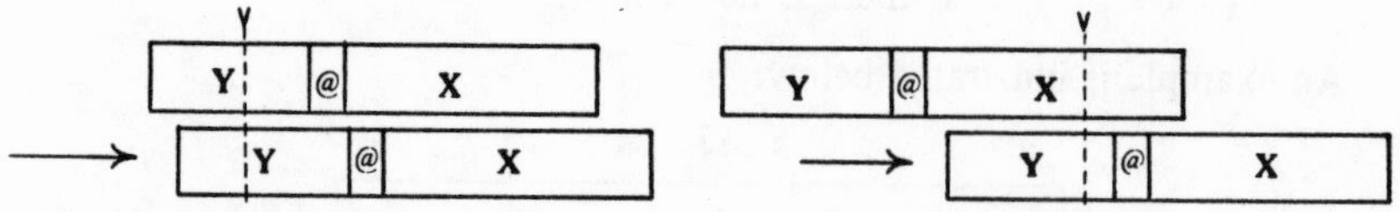

Preprocessing phase　　　　Output phase

3. A Turing maching implementation. The linear time bound obtained in §2 for the Morris-Knuth-Pratt algorithm seems to depend not only on the use of a random access machine, but also on the assignment of unit cost to a memory access, for just the P array alone contains $O(n \log n)$ bits when represented as a sequence of binary integers. This makes a linear time Turing maching implementation somewhat surprising.

The central economy results from representing the P array by a table Δ of differences. Define $P(-1) = -1$ and let

$$\Delta(i) = 1 + P(i) - P(i+1), \quad -1 \leqslant i < n.$$

Then

$$P(i) = i - \sum_{j=-1}^{i-1} \Delta(j) \quad \text{and} \quad \sum_{j=-1}^{n-1} \Delta(j) = n - P(n) \leqslant n + 1,$$

so the Δ array can be represented in linear space, even using unary notation.

We may expand the recursive definition of P in §2 as follows:

ALGORITHM X.

Stage (0). Set $P(0) \leftarrow -1$. Go to stage (1, 1).

Stage $(i+1, k)$.

1. If $Z_{[P^{(k)}(i)+1]} = Z_{i+1}$, set $P(i+1) \leftarrow P^{(k)}(i) + 1$ and go to stage $(i+2, 1)$.
2. If $P^{(k)}(i) = -1$, set $P(i+1) \leftarrow -1$ and go to stage $(i+2, 1)$.
3. Otherwise, go to stage $(i+1, k+1)$.

Algorithm X may be rewritten without explicit reference to P by using the Δ array and three new variables p, s and d. Inductively, at the beginning of stage $(i+1, k)$, the variables will satisfy

$$p = P^{(k)}(i), \tag{3.1}$$

$$s = P^{(k)}(i) - P^{(k+1)}(i), \tag{3.2}$$

$$d = P(i) - P^{(k)}(i). \tag{3.3}$$

Algorithm Y maintains these conditions and computes the Δ array. (The column vector notation denotes simultaneous assignment.)

ALGORITHM Y.

Stage (0).

$\Delta(-1) \leftarrow 1;$

$$\begin{bmatrix} p \\ s \\ d \end{bmatrix} \leftarrow \begin{bmatrix} -1 \\ 0 \\ 0 \end{bmatrix};$$

go to stage (1,1).

Stage $(i+1, k)$.

1. If $Z_{p+1} = Z_{i+1}$, then begin $\Delta(i) \leftarrow d;$

$$\begin{bmatrix} p \\ s \\ d \end{bmatrix} \leftarrow \begin{bmatrix} p+1 \\ s+\Delta(p) \\ 0 \end{bmatrix};$$

go to stage $(i+2, 1)$; end.

2. If $p = -1$, then begin $\Delta(i) \leftarrow d + 1;$

$$\begin{bmatrix} p \\ s \\ d \end{bmatrix} \leftarrow \begin{bmatrix} p \\ s \\ 0 \end{bmatrix};$$

go to stage $(i+2, 1)$; end.

3. Otherwise, begin

$$\begin{bmatrix} p \\ s \\ d \end{bmatrix} \leftarrow \begin{bmatrix} p-s \\ s - \sum_{j=p-s}^{p-1} \Delta(j) \\ d+s \end{bmatrix};$$

go to stage $(i+1, k+1)$; end.

It may be readily verified that conditions (3.1)–(3.3) hold after stage (0), are preserved by the remaining stages, and the Δ's are computed correctly.

The Turing machine to implement Algorithm Y has four tapes, each one-way infinite to the right. The input tape Z has two heads A and B. Tape Y has two heads C and D and holds the Δ's and d. p is represented by the positions of heads A and C. Tape S is used as a counter and holds s. Tape T is a scratch tape. The tapes with two heads may be replaced without time loss by several tapes with only one head per tape [3].

At the start of stage $(i+1, k)$, head A is scanning Z_p and head B is scanning Z_i. Tape Y contains the binary word

$$01^{\Delta(-1)}01^{\Delta(0)}01^{\Delta(1)}0 \cdots 01^{\Delta(i-1)}01^{d},$$

head C is on the "0" immediately preceding the block $1^{\Delta(p)}$ (the $(p+2)$nd "0" from the left), and head D is on the right-most nonblank square. Finally, the counter S contains the number s.

Below are the Turing machine tapes of the example of §2 at the beginning of stage (6, 2).

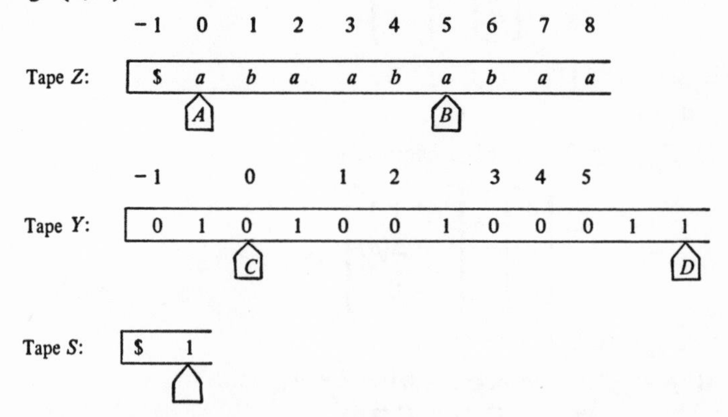

We now examine the operations that might be required in stage $(i+1, k)$. The first test, "$Z_{p+1} = Z_{i+1}$", is accomplished in three Turing machine steps, for heads A and B are only one square away from the symbols Z_{p+1} and Z_{i+1}, respectively. Similarly, the test "$p = -1$" becomes a test to see if head A is scanning the left endmarker, "$".

The updating in case (1) is accomplished by shifting D right and printing a "0", shifting heads A and B right one square, and moving C right to the next "0". As C is advanced, S is incremented once for each "1" that C passes over.

The updating in case (2) is even easier. A and C are left alone, B moves right one square, and D moves right two squares, printing a "1" followed by a "0".

To accomplish case (3), head C is moved left over s zeros. For each "0" passed over by C, head A moves one square to the left and head D moves one square to the right and prints a "1". For each "1" passed over by C, S is decremented. Since the counter S is modified by this process, its contents are first copied into the temporary counter T which is then used to control the iteration.

We total up separately the time spent in each of the three cases. For each i, case (2) is executed during at most one of the stages $(i+1, k)$. Each such execution takes a constant amount of time, so the total over all stages is clearly $O(n)$.

When case (3) is executed at stage $(i+1, k)$, it takes time cs for some constant c, where $s = P^{(k)}(i) - P^{(k+1)}(i)$ is the value of S at the start of the stage, for $\Sigma_{j=p-s}^{p-1} \Delta(j) \leqslant s$. Let k_i be the largest value of k for which a stage $(i+1, k)$ is executed. Then stages $(i+1, 1), \cdots, (i+1, k_i - 1)$ all execute case (3) and stage $(i+1, k_i)$ executes case (1) or (2). Hence, the total time spent in case (3) from the start of stage $(i+1, 1)$ to the start of stage $(i+2, 1)$ is

$$\sum_{k=1}^{k_i-1} c(P^{(k)}(i) - P^{(k+1)}(i)) = c(P(i) - P^{(k_i)}(i)) \leqslant c(P(i) - P(i+1) + 1).$$

Summing over all i, the total time in case (3) is $O(n)$.

Finally, the time spent in case (1) is bounded by the number of times C is shifted right. But this is at most the eventual length l_Y of tape Y plus the number of times C is shifted left. The latter occurs only in case (3) and hence is bounded by $O(n)$. Since

$$l_Y = n + 2 + \sum_{j=-1}^{n} \Delta(j) < 2n + 3 = O(n),$$

the total time spent in case (1) is also $O(n)$.

It follows that the total time of the Turing machine is $O(n)$.

4. "Don't care" symbols. An interesting extension of this simple string-matching problem which has practical applications results from the introduction of a "don't care" symbol, ϕ, into the alphabet. ϕ has the property of "matching" with any symbol. We shall write "$\equiv$" for this matching, so

$$\phi \equiv x \quad \text{for all } x \in \Sigma \cup \{\phi\}.$$

The Morris-Knuth-Pratt algorithm breaks down in this situation, basically because "$\equiv$" is not a transitive relation, that is

$$x \equiv y \wedge y \equiv z \not\rightarrow x \equiv z.$$

The above implication is valid only if $y \neq \phi$. Transitivity was assumed implicitly in deriving the recursive relation used to compute P. The naive algorithm given initially works just as before, with "$\equiv$" in place of "=." The ostensible aim of this paper is to produce a more efficient algorithm for string-matching with "don't care" symbols. This is achieved only for the case when Σ is a finite alphabet.

5. Generalised linear products. Both of the string-matching problems described so far can be regarded as special cases of a very general "linear product."

Given two vectors of elements, $\mathbf{X} = X_0, \cdots, X_m$ and $\mathbf{Y} = Y_0, \cdots, Y_n$, the *linear product with respect to* $\otimes$ *and* $\oplus$, written $\mathbf{X} \boxed{\substack{\otimes \\ \oplus}} \mathbf{Y}$, is a vector $\mathbf{Z} = Z_0, \cdots, Z_{m+n}$ where

$$Z_k = \bigoplus_{i+j=k} X_i \otimes Y_j \quad \text{for } k = 0, \cdots, m+n.$$

For this to be meaningful, $X_i, Y_j \in D$, $Z_k \in E$, for some sets D, E, and $\otimes$, $\oplus$ are functions

$$\otimes\colon D \times D \to E,$$

$$\oplus\colon E \times E \to E, \qquad \oplus \text{ associative.}$$

If $\oplus$ is $\wedge$, and $\otimes$ is $=$ or $\equiv$, the middle $m - n + 1$ truth values of the linear product give the information required in matching the text $\mathbf{X}$ against the *reversal* of $\mathbf{Y}$, that is $Y_n \cdots Y_0$, since

$$\left(\mathbf{X} \boxed{\substack{\equiv \\ \wedge}} \mathbf{Y}\right)_k = \mathit{true} \leftrightarrow [X_{k-n} \cdots X_k] \equiv [Y_n \cdots Y_0]$$

for $n \leqslant k \leqslant m$. The reason for introducing general linear products here lies in the following two cases.

(i) *Boolean product* where $\oplus$ is $\vee$ and $\otimes$ is $\wedge$, and

(ii) *polynomial product* where $\oplus$ is $+$ and $\otimes$ is $\times$.

The polynomial product is of course the ordinary multiplication of polynomials. The four products with which we are principally concerned are illustrated in Figure 1.

LINEAR PRODUCT $\mathbf{Z} = \mathbf{X} \boxed{\substack{\otimes \\ \oplus}} \mathbf{Y}$; $Z_k = \bigoplus_{i+j=k} X_i \otimes Y_j$.

EXAMPLES. With the convention that 1, 0 represent *true*, *false* respectively.

(1)

```
    b a a b a
      b a a      [≡/∧]
  -------------
    0 1 1 0 1 )
  0 1 1 0 1   }  ∧
1 0 0 1 0     )
  -------------
1 0 0 1 0 0 1
```

(2)

```
    a b φ φ a
      a φ b      [≡/∧]
  -------------
    0 1 1 1 0 )
  1 1 1 1 1   }  ∧
1 0 1 1 1     )
  -------------
1 0 0 1 1 1 0
```

(3)

```
      1 0 0 1 0 1 0 1
            1 0 1 0 1   [∧/∨]
  -------------------------
1 0 1 1 1 1 0 1 0 1 0 1
```

(4)

```
      1 0 0 1 0 1 0 1
            1 0 1 0 1   [×/+]
  -------------------------
1 0 1 1 1 2 0 3 0 2 0 1
```

FIGURE 1

6. Algorithms for linear products. For the simple string products the Morris-Knuth-Pratt algorithm can be extended to yield the complete linear product $Z_0 \cdots Z_{m+n}$. If we compute P for the string $\mathbf{Y}^R @ \mathbf{X}$, then the *middle* $m - n + 1$ digits of $\mathbf{Z}$ are given by:

$$\text{for } n \leqslant r \leqslant m, \quad Z_r = 1 \leftrightarrow P(r + n + 2) = n,$$

and the *last* n digits by:

for $m < r \leqslant m + n$, $Z_r = 1 \leftrightarrow P^{(k)}(m + n + 2) = n + m - r$ for some $k \geqslant 1$. For the *first* n digits, we know of no better method than to reverse both strings and use the same procedure.

For strings over a finite alphabet with "don't cares," we follow an indirect course, showing first that the computation time for string product is of the same order as that for Boolean product. If σ, τ are two distinct symbols of Σ, and $\mathbf{X}$ contains only σ's and ϕ's while $\mathbf{Y}$ contains only τ's and ϕ's, then the string product of $\mathbf{X}$ and $\mathbf{Y}$ is precisely the negation of the Boolean product of the strings $\hat{\mathbf{X}}$ and $\hat{\mathbf{Y}}$, where

$$\hat{X}_i = \mathit{true} = 1 \leftrightarrow X_i = \sigma, \qquad \hat{Y}_i = \mathit{true} = 1 \leftrightarrow Y_i = \tau$$

since

$$\bigwedge_{i+j=k} X_i \equiv Y_j \leftrightarrow \bigwedge_{i+j=k} \neg \hat{X}_i \vee \neg \hat{Y}_j \leftrightarrow \neg \bigvee_{i+j=k} \hat{X}_i \wedge \hat{Y}_j.$$

Thus Boolean product is no harder than ϕ-string product. On the other hand, let H_ρ be the predicate on $\Sigma \cup \{\phi\}$ defined by

$$\begin{aligned} H_\rho(x) &= 1 \quad \text{if } x = \rho, \\ &= 0 \quad \text{if } x \neq \rho \text{ (or } x = \phi), \end{aligned}$$

and extend H_ρ to strings in the obvious way. Then

$$\mathbf{Z} = \mathbf{X} \boxed{\begin{smallmatrix}\equiv \\ \wedge\end{smallmatrix}} \mathbf{Y} = \neg \bigvee_{\sigma \neq \tau; \sigma, \tau \in \Sigma} H_\sigma(\mathbf{X}) \boxed{\begin{smallmatrix}\wedge \\ \vee\end{smallmatrix}} H_\tau(\mathbf{Y}).$$

Informally, this equation states that $\mathbf{X}$ and $\mathbf{Y}^R$ match in a given relative position if and only if there is no pair of distinct symbols $\sigma, \tau \in \Sigma$ which clash. Hence the ϕ-string product takes the same time as the Boolean product to within a constant factor, independent of m and n.

There is a considerable similarity between the Boolean product and the polynomial product over the integers, as is shown in the example above. When 1 and 0 are identified with *true* and *false* respectively, the Boolean product can be obtained by performing the polynomial product and then by replacing any nonzero element by 1. This idea of embedding a Boolean algebra in a ring for computational purposes has been exploited to achieve a fast Boolean matrix multiplication and transitive closure algorithm [1].

One very convenient way to compute the polynomial product is to embed the product in a single large integer multiplication, for which there are a variety of well-known efficient algorithms. For the polynomial product of the $\{0, 1\}$-strings $X_0, \cdots, X_m$ and $Y_0, \cdots, Y_n$, where $m \geqslant n$, the maximum possible coefficient in the product is $n + 1$, If we choose r so that $2^r > n + 1$, compute the integers

$$X(2^r) = \sum_{i=0}^{m} X_i \cdot 2^{ri} \quad \text{and} \quad Y(2^r) = \sum_{j=0}^{n} Y_j \cdot 2^{rj}$$

and then multiply $X(2^r)$ by $Y(2^r)$, the result will be the product polynomial Z, evaluated at 2^r. Successive blocks of length r in the binary representation of $Z(2^r)$ will give the coefficients of Z, and by replacing nonzero coefficients by 1 we obtain the elements of the Boolean product. This is illustrated below.

$r > \log_2(n + 1),\ m \geqslant n$

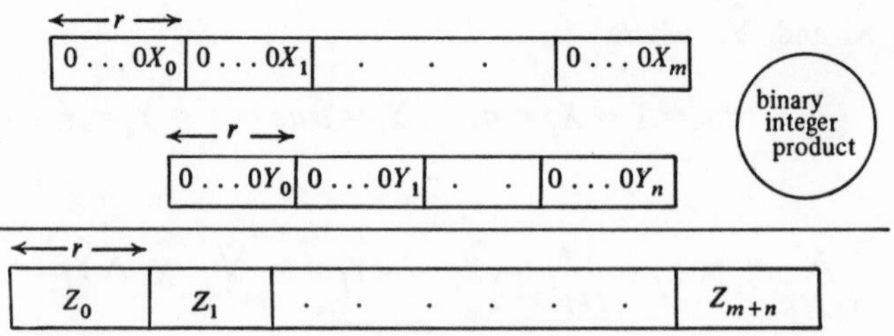

$\longleftarrow r \longrightarrow$

$0 \ldots 0X_0$	$0 \ldots 0X_1$	$\cdot \quad \cdot \quad \cdot$	$0 \ldots 0X_m$

$\longleftarrow r \longrightarrow$

$0 \ldots 0Y_0$	$0 \ldots 0Y_1$	$\cdot \quad \cdot$	$0 \ldots 0Y_n$

binary integer product

$\longleftarrow r \longrightarrow$

Z_0	Z_1	$\cdot \quad \cdot \quad \cdot \quad \cdot \quad \cdot \quad \cdot$	Z_{m+n}

where $Z = X \boxed{\substack{\times \\ +}} Y$. $\boxed{\substack{\times \\ +}}$ is polynomial product.

The operations required to construct $X(2^r)$ and $Y(2^r)$, and to pick out the coefficients of Z are very easy and efficient on a binary computer. On most computers there is fast special-purpose hardware for multiplication of integers up to a certain size, and efficient routines for multiplying larger integers. These may be used to yield a good practical program for the Boolean product of strings of moderate length, which however has a running time that is still proportional to to mn.

For truly large integers, the Schönhage-Strassen algorithm [5] multiplies M-digit numbers by N-digit numbers in a time which is $O(M \cdot \log N \cdot \log \log N)$ for $M \geqslant N$, using a multi-tape Turing machine. For our application, $M = mr = O(m \log n)$ and $N = nr = O(n \log n)$. Hence:

RESULT. For a finite alphabet, the "don't care" product of strings of lengths m and n $(m \geqslant n)$ can be computed with a multi-tape Turing machine in time $O(m \cdot (\log n)^2 \cdot \log \log n)$.

7. Large alphabets and numbers of comparisons. The algorithm for ϕ-product described so far has the disadvantage that the running time increases rapidly with the size of the alphabet Σ. It is approximately proportional to $|\Sigma|^2$. By coding the symbols of Σ into a binary alphabet we can use just two Boolean products for strings of length $m \cdot \log|\Sigma|$ and $n \cdot \log|\Sigma|$. Provided $|\Sigma|$ is bounded by a power of n, this introduces a factor of just $\log|\Sigma|$ into the running time.

It is interesting to observe that the Morris-Knuth-Pratt algorithm works for an infinite alphabet, provided we take the predicate "=" as a basic operation. Our algorithm for ϕ-product is not of this form and we may ask whether there is any algorithm, with access to the strings only through the predicate "$\equiv$", which has a computation time better than the obvious $O(m \cdot n)$. Under such a strict limitation the answer is "no", and this is easily seen by considering the product of the two strings $\mathbf{X} = \phi^{m+1}$ and $\mathbf{Y} = \phi^{n+1}$. All $\equiv$-tests have the result *true*, but suppose that during the execution of some algorithm there is some test "$X_i \equiv Y_j$?" which is never made. The computation and output would be indistinguishable from that for the pair of strings $\phi^i \sigma \phi^{m-i}$ and $\phi^j \tau \phi^{n-j}$, where $\sigma, \tau \in \Sigma$, and $\sigma \neq \tau$, and therefore the algorithm cannot correctly compute the string product. Hence any ϕ-product algorithm of this class must sometimes make at least $(m + 1)(n + 1)$ tests.

The above restriction is perhaps a little severe, even if we consider the case of infinite Σ; so let us allow in addition an explicit test for the "don't care" symbol, that is "$X_i = \phi$?" or "$Y_j = \phi$?". The lower bound on the number of tests is now radically different, for we can show that $O(m + n)$ are sufficient. Unfortunately we still know of no algorithm with a *total* running time less than that of the naive algorithm for the ϕ-product over an infinite alphabet.

8. Algorithm for ϕ-product using $O(m + n)$ tests. We have to evaluate the $(m + n + 1)$ conjunctions

$$Z_k = \bigwedge_{i+j=k} X_i \equiv Y_j \quad \text{for } k = 0, \cdots, m + n.$$

First we determine all occurrences of ϕ in $\mathbf{X}$ and $\mathbf{Y}$, and replace by *true* any equivalence involving ϕ. Possibly some of the Z_k's may thus be determined. So far as we know the remaining symbols in $\mathbf{X}$ and $\mathbf{Y}$ may be completely distinct. At each stage of the algorithm we shall maintain an equivalence relation on these symbols, such that we have determined that all symbols in the same equivalence class are identical. We can always choose "$X_i \equiv Y_j$?" for our next comparison, where X_i and Y_j are in distinct equivalence classes, and Z_{i+j} has yet to be determined. If $X_i \equiv Y_j$, then the equivalence classes of X_i and Y_j can be united, whereas if $X_i \not\equiv Y_j$ then Z_{i+j} can be determined as *false*. If during the course of the algorithm the former case occurs $(m + n + 1)$ times then only one equivalence class remains, and if the latter case occurs $(m+n+1)$ times then all the Z_k's have been determined. Either way, no further comparisons are required. Hence at most $2(m+n+1)$ equality tests and $m+n+2$ ϕ-tests are needed, giving a total which is $O(m+n)$.

9. On-line palindromes. A computation is performed *on-line* if the ith output symbol is produced before the $(i+1)$st input symbol is read. Let $Z_i = 1$ if $X_0 \cdots X_i$ is a palindrome (i.e. if $X_0 \cdots X_i = X_i \cdots X_0$), and $Z_i = 0$ otherwise. Then

$$\mathbf{Z} = \mathbf{X} \boxed{\overset{=}{\wedge}} \mathbf{X};$$

so $\mathbf{Z}$ can be computed in time $O(n)$, even on a Turing machine, as outlined in §§3 and 6 using the Morris-Knuth-Pratt algorithm

Fischer and Stockmeyer [2] present a general procedure for converting any off-line multiplication algorithm which runs in time $T(n)$ to an on-line method taking time $O(T(n) \log n)$ when T satisfies $T(2n) \geqslant 2T(n)$. Their construction applies to any generalised linear product; so, in particular, the time $O(n)$ method above for computing $\mathbf{Z}$ can be converted to an on-line Turing machine program that runs in time $O(n \log n)$.[1]

10. Conclusions and open problems. We have considered string-matching problems with and without a "don't care" symbol. In both cases a naive procedure, based directly on the problem definition, takes time proportional to $m \times n$ where m, n are the lengths of the two strings to be matched. The Morris-Knuth-Pratt algorithm provides a practical and elegant way to compute the former problem in $O(m+n)$ time, but there seems to be no obvious extension of their algorithm to the "don't care" case. This is partially explained by our lower bound result which shows that $O(mn)$ is the best possible bound unless more information is allowed than the mere results of comparisons between pairs of symbols. With a further basic test which explicitly detects the "don't care" symbol, this lower bound collapses and there is at least the possiblity of a faster algorithm. Provided that the symbol alphabet is finite, we have demonstrated an algorithm with a running time which is $O(m \cdot \log n \cdot \log \log n)$. The method is indirect and not of practical value except for very large m and n; however it shows the feasibility of algorithms which are faster than the naive procedure for "don't care" matching.

We have not treated at all the superficially similar problem, where a "don't

[1] *Added in proof.* Slisenko has announced a method for recognizing palindromes in real-time on a multi-tape Turing machine [6].

care" symbol can "match" an arbitrary *string* of symbols. A good algorithm for this would have obvious practical applications.

We have only begun to compare and contrast the computational complexity of generalised linear products for various ⊗ and ⊕. There are several more, interesting, structures for which the linear product is a natural operation. A study of algorithms for linear products, based on the axiomatic properties of ⊗ and ⊕, may provide valuable insight into why some products are easier than others.

References

1. M. J. Fischer and A. R. Meyer, *Boolean matrix multiplication and transitive closure*, 12th IEEE Sympos. on Switching and Automata Theory, 1971, pp. 129–131.

2. M. J. Fischer and L. J. Stockmeyer, *Fast on-line integer multiplication*, 5th ACM Sympos. on Theory of Computing, 1973, pp. 67–72; J. Comput. Sys. Sci. 9 (1974) (to appear).

3. P. C. Fischer, A. R. Meyer and A. L. Rosenberg, *Real-time simulation of multihead tape units*, J. ACM 19 (1972), 590–607.

4. J. H. Morris and V. R. Pratt, *A linear pattern-matching algorithm*, TR-40, Computer Center, Univ. of Calif., Berkeley (June 1970).

5. A. Schönhage and V. Strassen, *Schnelle Multiplikation grosser Zahlen*, Computing (Arch. Elektron. Rechnen) 7 (1971), 281–292. MR 45 #1431.

6. A. O. Slisenko, *Recognition of palindromes by multihead Turing machines*, Proc. Steklov Math. Inst. Acad. Sci. USSR 129 (1973), 30–202.

MASSACHUSETS INSTITUTE OF TECHNOLOGY

UNIVERSITY OF WARWICK

SIAM-AMS Proceedings
Volume 7
1974

The Evaluation of Determinants by Expansion by Minors and the General Problem of Substitution*

W. M. Gentleman and S. C. Johnson

Motivation. One of the least understood aspects of symbolic formula manipulation is the problem of substitution, particularly the merits of various strategies. A specific example where substitution is complicated enough to be interesting, and where theoretical results can be obtained, is the problem of evaluating a determinant by substitution into the Laplace expansion.

Introduction. One obvious possibility for evaluating determinants is substitution into the Laplace expansion, regarded as a polynomial form of $n!$ terms in n^2 indeterminates. The obvious way to substitute into this form requires $(n - 1)$ multiplications per term, a total of $(n - 1)n!$ multiplications. Many of these multiplications are redundant, and no advantage is taken of any collection of like terms. Some sort of parenthesizing and grouping of terms, as in Horner's rule, is clearly indicated.

Many such groupings are possible. For example, when $n = 4$, the $4!$ terms

$$\begin{aligned}
&x_{11}x_{22}x_{33}x_{44} - x_{11}x_{22}x_{34}x_{43} - x_{11}x_{23}x_{32}x_{44} + x_{11}x_{23}x_{34}x_{42} \\
&- x_{11}x_{24}x_{33}x_{42} + x_{11}x_{24}x_{32}x_{43} - x_{12}x_{21}x_{33}x_{44} + x_{12}x_{21}x_{34}x_{43} \\
&+ x_{12}x_{23}x_{31}x_{44} - x_{12}x_{23}x_{34}x_{41} + x_{12}x_{24}x_{33}x_{41} - x_{12}x_{24}x_{31}x_{43} \\
&+ x_{13}x_{21}x_{32}x_{44} - x_{13}x_{21}x_{34}x_{42} - x_{13}x_{22}x_{31}x_{44} + x_{13}x_{22}x_{34}x_{41} \\
&+ x_{13}x_{24}x_{31}x_{42} - x_{13}x_{24}x_{32}x_{41} + x_{14}x_{21}x_{33}x_{42} - x_{14}x_{21}x_{32}x_{43} \\
&+ x_{14}x_{22}x_{31}x_{43} - x_{14}x_{22}x_{33}x_{41} - x_{14}x_{23}x_{31}x_{42} + x_{14}x_{23}x_{32}x_{41}
\end{aligned}$$

AMS (MOS) subject classifications (1970). Primary 68A15, 68A20.

*This article appeared in Math. Comp. 28 (1974), 543–548.

can be written, among other ways, as

$$
\begin{aligned}
&(x_{11}x_{22} - x_{12}x_{21})(x_{33}x_{44} - x_{34}x_{43})\\
&- (x_{11}x_{23} - x_{13}x_{21})(x_{32}x_{44} - x_{34}x_{42})\\
&+ (x_{11}x_{24} - x_{14}x_{21})(x_{32}x_{43} - x_{33}x_{42})\\
&+ (x_{12}x_{23} - x_{13}x_{22})(x_{31}x_{44} - x_{34}x_{41})\\
&- (x_{12}x_{24} - x_{14}x_{22})(x_{31}x_{43} - x_{33}x_{41})\\
&+ (x_{13}x_{24} - x_{14}x_{23})(x_{31}x_{42} - x_{32}x_{41})
\end{aligned}
$$

or again as

$$
\begin{aligned}
&(a_1x_{33} - a_2x_{32} + a_4x_{31})x_{44}\\
&- (a_1x_{34} - a_3x_{32} + a_5x_{31})x_{43}\\
&+ (a_2x_{34} - a_3x_{33} + a_6x_{31})x_{42}\\
&- (a_4x_{34} - a_5x_{33} + a_6x_{32})x_{41}
\end{aligned}
$$

where

$$
\begin{aligned}
a_1 &= (x_{11}x_{22} - x_{12}x_{21}), & a_2 &= (x_{11}x_{23} - x_{13}x_{21}),\\
a_3 &= (x_{11}x_{24} - x_{14}x_{21}), & a_4 &= (x_{12}x_{23} - x_{13}x_{22}),\\
a_5 &= (x_{12}x_{24} - x_{14}x_{22}), & a_6 &= (x_{13}x_{24} - x_{14}x_{23}).
\end{aligned}
$$

When such groupings are tried, the productive ones simply turn out to be statements or the well-known general combinatorial definition of determinants, expanding by minors:

DEFINITION. The determinant of a square matrix of order 1 is the entry in that matrix. The determinant of a matrix of order n can be found by choosing m less than n, and selecting m columns from the original determinant. Consider then each of the $\binom{n}{m}$ ways that m rows can be chosen from the original determinant. Each row choice defines two submatrices: one of order m corresponding to the chosen rows and columns, and one of order $n - m$ corresponding to the complementary sets. The original determinant is the sum (with appropriate signs), over all row choices, of the product of these two smaller determinants (minors).

Any direct implementation of this recursive definition is very inefficient, as it leads to the repeated recalculation of small minors. Nonrecursive implementations, however, can be obtained in the following manner: Consider any rooted binary (bifurcating) tree with n leaves labelled 1 through n (for example, Figures 1 and 2 below). With each node in the tree (including the leaves), we associate a set of minors, as follows: If the subtree rooted at the given node has m leaves, we associate with this node the set of $\binom{n}{m}$ minors obtained by all possible choices of m rows, with the column indices given by the m leaf names.

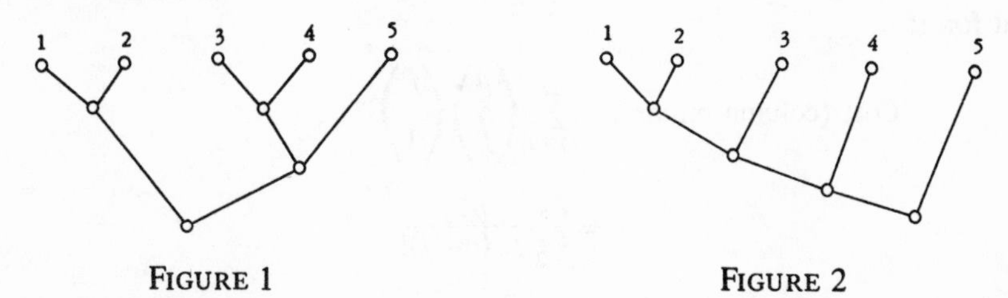

FIGURE 1 FIGURE 2

We note the following simple properties of these minor sets:

(1) The minor set for the root of the tree has a single element, the determinant.

(2) The minor set of leaf k has m elements, which are simply the elements of column k of the matrix.

(3) Using the above combinatorial definition of determinants we can compute the minor set of a given node from the minor sets of its two immediate descendants. In this way, every tree corresponds to a (nonrecursive) algorithm for computing the determinant by minor expansion.

As a specific example, consider the classical method of expansion by column minors. In this method, one computes all the minors determined by two particular columns, then uses these to obtain all the minors determined by these columns plus one additional column, then uses these to obtain all the minors determined by four columns, etc. This corresponds to a tree, such as in Figure 2, where each node except leaves has one descendant which is a leaf. In fact, the following theorem shows that expansion by column minors is one of the best minor expansions.

THEOREM. *The minimum cost (in multiplications) of expanding an* $n \times n$ *determinant by minors (using any of the above trees) is* $n(2^{n-1} - 1)$, *which is the cost of expanding by column minors.*

PROOF. Suppose that the subtree rooted at a specific node has m leaves, and its two descendant subtrees have k and $m - k$ leaves, respectively, where $k \leqslant m - k$. There are $\binom{n}{m}$ minors of order m in the minor set of this node;

each is obtained by forming and summing $\binom{m}{k}$ products of minors from the minor sets of the two descendant nodes. The cost (in multiplications) of producing all $\binom{n}{m}$ minors is thus $\binom{n}{m}\binom{m}{k}$, plus the cost of obtaining the minors defined by each subtree.

The minor set of this node could obviously be computed by expansion by column minors. From the preceding remarks this would cost $\Sigma_{j=2}^{m}\binom{n}{j}\binom{j}{1}$ (since $k = 1$ at each of the $m - 1$ stages). For $m = n$, i.e., for computing the original determinant by expansion by column minors, this can be summed in closed form:

$$\text{Cost (column minors)} = \sum_{j=2}^{n}\binom{n}{j}\binom{j}{1}$$

$$= \sum_{j=2}^{n}\frac{n!\,j}{j!\,(n-j)!}$$

$$= n\sum_{j=2}^{n}\frac{(n-1)!}{(j-1)!\,(n-j)!}$$

$$= n\sum_{j=1}^{n-1}\binom{n-1}{j} = n(2^{n-1} - 1).$$

We will prove this is minimal over all binary trees by inductively showing that to compute the minor set of a node covering m leaves, the minimal cost is $\Sigma_{j=2}^{m}\binom{n}{j}j$. This is clearly true if $m = 2$ or even $m = 3$, since there only is one possible tree. Assume it is true for $2, 3, \cdots, m-1$. The minimum cost to cover m columns is this, for some k:

$$\binom{n}{m}\binom{m}{k} + \sum_{j=2}^{k}\binom{n}{j}j + \sum_{j=2}^{m-k}\binom{n}{j}j.$$

Our theorem is proved if we show that this cost is minimized at $k = 1$. The difference between the cost for k and the cost for $k - 1$ is

$$\binom{n}{m}\binom{m}{k} - \binom{n}{m}\binom{m}{k-1} + \binom{n}{k}k - \binom{n}{m-k+1}(m-k+1).$$

We will show that this is nonnegative for $k \geqslant 2$, thus $k = 1$ is minimal. Rewriting the above expression, what we want to prove is

$$\binom{n}{m}\left[\binom{m}{k} - \binom{m}{k-1}\right] + \binom{n}{k}k - \binom{n}{m-k+1}(m-k+1) \geqslant 0,$$

i.e.,

$$\frac{n!}{m!\,(n-m)!}\left[\frac{m!}{k!\,(m-k)!}-\frac{m!}{(k-1)\,!\,(m-k+1)!}\right]$$

$$+\frac{n!\;k}{k!\,(n-k)!}-\frac{n!\,(m-k+1)}{(m-k+1)!\,(n-m+k-1)!}$$

$$=\frac{n!}{(n-m)!\,(k-1)!\,(m-k)!}$$

$$\cdot\left[\frac{1}{k}-\frac{1}{m-k+1}+\frac{(n-m)!\,(m-k)!}{(n-k)!}-\frac{(n-m)!\,(k-1)!}{(n-m+k-1)!}\right]\geqslant 0$$

and since the factor outside the square brackets is clearly positive, we need only consider the quantity within, which can be written as

$$\frac{m-2k+1}{k(m-k+1)}+\binom{n-k}{n-m}^{-1}-\binom{n-m+k-1}{n-m}^{-1}.$$

It is convenient to rewrite this in terms of $s=n-m$ and $t=m-2k$ as

$$D(k,s,t)=\frac{t+1}{k(k+t+1)}+\binom{k+t+s}{s}^{-1}-\binom{k+s-1}{s}^{-1}$$

which we must prove nonnegative for $k\geqslant 2,\ s\geqslant 0,\ t\geqslant 0$. We will do this by considering five cases.

Case I. $s=0, t\geqslant 0, k\geqslant 2.$

$$D(k,s,t)=\frac{t+1}{k(k+t+1)}+1-1=\frac{t+1}{k(k+t+1)}$$

which is clearly positive.

Case II. $s=1, t\geqslant 0, k\geqslant 2.$

$$D(k,s,t)=\frac{t+1}{k(k+t+1)}+\frac{1}{k+t+1}-\frac{1}{k}=0.$$

Case III. $s=2, t\geqslant 0, k\geqslant 2.$

$$D(k,s,t)=\frac{t+1}{k(k+t+1)}+\binom{k+t+2}{2}^{-1}-\binom{k+1}{2}^{-1}$$

$$=\frac{t+1}{k(k+t+1)}+\frac{2}{(k+t+1)(k+t+2)}-\frac{2}{k(k+1)}$$

$$=\frac{(k-1)t^2+(k^2-3)t+(k-2)(k+1)}{k(k+1)(k+t+1)(k+t+2)}$$

and since this is a polynomial in t which, for any $k \geqslant 2$, has nonnegative coefficients, it is nonnegative for $t \geqslant 0$.

Case IV. $s \geqslant 3, t = 0, k \geqslant 2$.

$$\begin{aligned} D(k, s, t) &= \frac{1}{k(k+1)} + \binom{k+s}{s}^{-1} - \binom{k+s-1}{s}^{-1} \\ &= \frac{1}{k(k+1)} + \frac{k!\, s!}{(k+s)!} - \frac{(k-1)!\, s!}{(k+s-1)!} \\ &= \frac{1}{k(k+1)} + \frac{k!\, s!}{(k+s)!}\left[1 - \frac{k+s}{k}\right] = \frac{1}{k(k+1)} - \frac{s\, k!\, s!}{k(k+s)!} \\ &= \frac{s^2}{k(k+1)}\left[\frac{1}{s^2} - \frac{(k+1)!\; (s-1)!}{(k+s)!}\right]. \end{aligned}$$

Thus it suffices to show $\binom{k+s}{k+1} \geqslant s^2$. But this follows since

$$\binom{k+s}{k+1} - s^2 \geqslant \binom{s+2}{3} - s^2 = \frac{(s+2)(s+1)s - 6s^2}{6} = \frac{s(s-1)(s-2)}{6}$$

which is nonnegative for $s \geqslant 2$.

Case V. $s \geqslant 3, t \geqslant 1, k \geqslant 2$.

$$D(k, s, t) \geqslant \frac{2}{k(k+2)} - \binom{k+s-1}{s}^{-1} \geqslant \frac{2}{k(k+2)} - \binom{k+2}{3}^{-1}.$$

But

$$\frac{2}{k(k+2)} - \binom{k+2}{3}^{-1} = \frac{2}{k(k+2)} - \frac{6}{k(k+1)(k+2)} = \frac{2}{k(k+2)}\left[1 - \frac{3}{k+1}\right]$$

which for $k \geqslant 2$ is nonnegative. Q. E. D.

Note. The above proof shows expanding by column minors to be optimal, but it is not the only optimal scheme. The order in which the columns are introduced is arbitrary, and, moreover, the role of columns and rows can obviously be interchanged.

Conclusions. At first glance, the theorem above seems rather negative: Everyone knows that expansion by column minors is too expensive to be a practical algorithm, and we have shown no other form of minor expansion is better. However, the well-known expense of minor expansion is an asymptotic statement, and for small n (less than 6) minor expansion is actually cheaper than Gaussian elimination. Moreover, the assertion about expense is based on an operation count that assumes all multiplications are equally expensive: In computations of symbolic algebra this is far from true, and determinant evaluation by minors is often the best way. Thus the study of minor expansion as a practical algorithm is relevant, and in this context the result is pleasantly surprising–surprising, because we usually expect algorithms associated with balanced trees to be preferable to those associated with unbalanced ones, whereas here the most unbalanced tree is best; and pleasant because expansion by column minors is certainly the simplest minor expansion to implement. Moreover, we get an extra benefit: The implication of the results on polynomial powering [1] is that it is generally better, given the choice, to perform operations between one derived quantity and one quantity from the original problem rather than to perform operations between two derived quantities, and of course that is what distinguishes expansion by column minors. (The advantages can be quantified here too, by only slight changes in the proof above.)

What does all this have to do with the general problem of substitution into an arbitrary form? In determinant evaluation we fortunately had a set of groupings that we could study theoretically. In the case of an arbitrary form, it is not clear what strategy to use in grouping terms. Horner's rule is satisfactory if the form is a univariate polynomial, but if the form is multivariate, the obvious recursive generalization of Horner's rule (regarding polynomials in $n-j$ variables as polynomials in 1 variable with coefficients that are polynomials in $n-j-1$ variables) is unsatisfactory since it can involve recomputing common subexpressions. On the other hand, it is most unlikely that any practical strategy can be more complicated than augmenting the obvious recursive scheme with a pattern match to recognize common subexpressions. Such strategy has the advantage for symbolic computation referred to above, i.e., that all operations are performed between a derived quantity and one from the original problem, but what other assurance have we that it is reasonable? One such is that this strategy, applied to the Laplace expansion, leads to what is effectively expansion by column minors.

References

1. W. M. Gentleman, *Optimal multiplication chains for computing a power of a symbolic polynomial,* Math. Comp. **26** (1972), 935–939. MR **47** #2855.

2. Sir Thomas Muir, *The theory of determinants in the historical order of development,* MacMillan, London, 1906–1920.

UNIVERSITY OF WATERLOO

SIAM-AMS Proceedings
Volume 7
1974

The Complexity of Linear Approximation Algorithms*

Martin H. Schultz

1. Introduction. The computation of simple approximations to general functions or data is a very common activity at most computing centers. In this paper, we discuss the complexity of such computations. We discuss questions which are direct analogues of those currently being duscussed in "concrete" complexity and computational combinatorics. Specifically we will concentrate on four themes: (1) the general problem is computationally difficult; (2) adaptive or artificial intelligence algorithms are computationally difficult; (3) subproblems are computationally easy; and (4) computationally easy "approximate" algorithms exist.

In this regard, the emphasis of this paper is somewhat different from that of the recent paper of J. Rice [14] on a similar topic and the work of J. Traub and others on "analytic computational complexity" (cf. [18]).

In §2, we show that for all reasonable mathematical models, linear approximation algorithms have infinite computational complexity (for the worst case analysis). Moreover, we show that nonlinear, adaptive algorithms are of no assistance in the worst case.

In §3, we derive *lower* and *upper* bounds for the error in approximating an important class of smooth functions defined on the unit interval [0, 1]. For the class of functions under consideration, we show that the subspace of continuous, piecewise linear polynomials with n uniformly spaced knots is an essentially optimal n-dimensional subspace. This demonstrates theme (3).

In §§4 and 5, we concentrate on theme (4) and study computationally easy, approximate mappings into subspaces of continuous, piecewise linear

AMS (MOS) subject classifications (1970). Primary 68A20, 65D10, 65D05.

* This research was supported in part by the Office of Naval Research, N0014-67-A-0097-0016.

polynomials. In §4, we introduce and study the mapping which yields a discrete Tchebycheff approximation and in §5 we consider the familiar interpolation and least squares projection mappings. Finally, in §6, we consider the extension of the material of §5 to the approximation of functions of two variables defined on a square domain.

Most of the results of this paper can be extended to subspaces of piecewise polynomials of arbitrary degree. What remains is a verification of many technical details many of which have already been provided by deBoor (cf. [2] and [3]). However, our goal in this paper is to present a point of view rather than mathematical generality and virtuosity. Hence, we consider only the technically simple case of piecewise linear polynomials.

2. A discouraging complexity result. The general mathematical framework for our study of linear approximation algorithms will be infinite-dimensional real Banach space B, i.e., an infinite-dimensional complete, normed, vector space over the real field. Our prime example will be the space of all real-valued, continuous functions f defined on the unit interval $[0, 1]$ with the maximum norm $\|f\| \equiv \max\{|f(x)| \mid 0 \leqslant x \leqslant 1\}$.

If S is an index set, an algorithm for linear approximation in B consists of a set of finite-dimensional subspaces of B, $\{B(s) | \, s \in S\}$, and a set of associated mappings $\{M(s) | \, s \in S\}$ such that $M(s)\colon B \rightarrow B(s)$. Our prime example of S will be the set of all ordered n-tuples $\Delta\colon 0 = x_1 < x_2 < \cdots < x_n = 1$ and our prime example of $B(s)$ will be the n-dimensional space $L(\Delta) \equiv \{l(x) \in C[0, 1] \mid l(x)$ is a linear polynomial on each element $[x_i, x_{i+1}]$, $1 \leqslant i \leqslant n - 1\}$, where $n \geqslant 2$.

We input as data to the algorithm the element $s \in S$ and the element $b \in B$. As output, we obtain $M(s)b$ which we hope is a "good" approximation. By "good" we mean that the error $E(b, s) \equiv \|b - M(s)b\|$ is sufficiently small.

Generally we are given a tolerance $\epsilon > 0$ and we must select s to guarantee that

$$E(b, s) \leqslant \epsilon. \tag{2.1}$$

For a *worst case analysis*, we wish to have (2.1) for all $b \in B$. Hence, we want

$$E(s) \equiv \sup\{E(b, s) \mid b \in B, \|b\| = 1\} \leqslant \epsilon. \tag{2.2}$$

Clearly we may view ϵ^{-1} as a fairly accurate parameterization of the computational difficulty of the approximation problem. If $\dim B(s) > \dim B(t)$ for $s, t \in S$, we expect that $E(s) < E(t)$ and $\operatorname{cost} M(s) > \operatorname{cost} M(t)$.

Recent work in complexity theory leads us to investigate the dependence of $E(s)$ on the dimension of $B(s)$. In particular, we would like to find a space

B and a linear approximation algorithm such that $E(s) = O((\dim B(s))^{-t})$ as $\dim B(s) \to \infty$, thus giving polynomial complexity. However, it may be that the best we can do is $E(s) = O((\log \dim B(s))^{-1})$ as $\dim B(s) \to \infty$, thus giving exponential complexity.

Unfortunately this problem is a disaster; it has infinite complexity! We will show that we always have $E(s) = 1$. This will show that no matter how clever we are a priori and no matter how much computer time we invest, there will be inputs for which our algorithm computes approximations which are no better than the zero of the Banach space.

THEOREM 2.1. *For all* B, $\{B(s) \mid s \in S\}$, *and* $s \in S$,

$$E(s) \equiv \sup\{\|b - M(s)b\| \mid b \in B, \|b\| = 1\} = 1. \tag{2.3}$$

PROOF. It suffices to show that for all $s \in S$, there exists $b \in B$ such that $\|b\| = 1$ and $E(s, b) = 1$. Since $B(s)$ is a closed, proper subspace of B, there exists a vector $y \notin B(s)$. If z denotes a best approximation to y in $B(s)$, then the vector $b \equiv (y - z)/\|y - z\|$ has the necessary properties. Q.E.D.

There are a number of valuable lessons to be learned from this result. First, we must take our input data from dense, nonclosed subspaces of Banach spaces. Second, worst case analyses may be misleading – after all, people do successfully use linear approximation algorithms in practical situations.

We might hope to rescued from our difficulties by resorting to nonlinear, adaptive algorithms. However, we will show that these approaches will not help as far as a worst case analysis goes.

We can model nonlinear, adaptive algorithms by assuming the algorithm "chooses" both $s \in S$ and $Mb \in B(s)$. For example, we may consider $L(\Delta)$, where Δ is a set of n knots, and allow our algorithm to vary the $n - 2$ internal knots.

However, under reasonable conditions (which are satisfied in the above example), we can prove an analogue of Theorem 2.1.

THEOREM 2.2. *If the closure of* $B(S) \equiv \bigcup_{s \in S} B(s)$ *is a proper subset of* B *and* M: $B \to B(S)$, *then*

$$E(S) \equiv \sup\{\|b - Mb\| \mid b \in B, \|b\| = 1\} = 1. \tag{2.4}$$

PROOF. It suffices to show that there exists $b \in B$ with $\|b\| = 1$ and

$$d(b, B(S)) \equiv \inf\{\|b - y\| \mid y \in B(s), s \in S\} = 1. \tag{2.5}$$

Assume (2.5) is false, i.e., there exists $\delta < 1$ such that

$$d(b, B(S)) \leqslant \delta \tag{2.6}$$

for all $b \in B$ with $\|b\| = 1$. Since the closure of $B(S)$, $\mathrm{cl}(B(S))$, is a proper subset of B, there exists $y \notin \text{cl}(B(S))$. Let $\{y_k\}_{k=1}^{\infty} \subset B(S)$ be such that $\|y - y_k\| \to d(y, B(S))$ as $k \to \infty$.

The sequence of vectors $v_k \equiv (y - y_k)/\|y - y_k\|$, $k \geq 1$, has $\|v_k\| = 1$ and, by (2.6),

$$d(v_k, B(S)) \leq \delta. \tag{2.7}$$

Thus, $d(y - y_k, B(S)) \leq \delta \|y - y_k\|$. Hence, $d(y, B(S)) = d(y - y_k, B(S)) \leq \delta \|y - y_k\|$ and taking the limit as $k \to \infty$, we obtain $d(y, B(S)) \leq \delta d(y, B(S)) < d(y, B(S))$, which is a contradiction. Q.E.D.

Thus, we have shown that, for a worst case analysis, nonlinear adaptive algorithms do not help us. Of course, for particular classes of problems they are very effective (cf. [13]).

3. Lower bounds. The results of §2 suggest that we should restrict our inputs f to our approximation algorithm if we hope to achieve some reasonable results. In 1936, the Russian mathematician Kolmogorov (cf. [8]) had the brilliant idea of studying the quantities

$$d_n(A) \equiv \inf_{B_n} \sup_{b \in A} \inf_{b_n \in B_n} \|b - b_n\|, \tag{3.1}$$

where A is the set of allowable inputs, n is a positive integer, and the infimum is over all n-dimensional subspaces B_n of B. Once we know the quantities $d_n(A)$, we have a hold on lower bounds on the complexity of linear approximation algorithms.

For the remainder of this paper we will restrict ourselves to the special case of $B \equiv C[0, 1]$ with norm $\|f\| \equiv \max\{|f(x)| \mid 0 \leq x \leq 1\}$ and $A \equiv \{f \mid f \in W^{1,\infty}(0, 1)$ and $\|Df\| \leq 1\}$, i.e., A is the set of absolutely continuous functions f with $\|Df\| \leq 1$. Following a technique given in [9], we may prove a lower bound due to Tihomirov [17].

THEOREM 3.1. $d_n(A) \geq 1/2n$.

PROOF. Let $B_n \subset C[0, 1]$ be any n-dimensional subspace spanned by $\phi_1(x), \cdots, \phi_n(x)$ and Δ_{n+1}: $0 = x_1 < x_2 < \cdots < x_{n+1} = 1$ be the uniform partition with uniformly spaced knots, $x \equiv (i - 1)/n$, $1 \leq i \leq n + 1$. If A is the $n \times (n + 1)$ matrix given by $A \equiv [a_{ij}] \equiv [\phi_i(x_j)]$, the linear system

$$A\mathbf{c} = \mathbf{0} \tag{3.2}$$

has a nontrivial solution $\tilde{\mathbf{c}}$ such that $\Sigma_{i=1}^{n+1} |\tilde{c}_i| = 1$.

If $\lambda_i \equiv \text{sign}\, c_i$, $1 \leq i \leq n + 1$, choose $l(x) \in L(\Delta_{n+1})$ such that sign $l(x_i) \equiv \lambda_i$, $1 \leq i \leq n + 1$, and $|l(x_i)| \equiv 1/2n$. Clearly $l(x) \in A$.

Moreover, for all $\mathbf{a} \in R^n$, we have

$$\left\| l - \sum_{k=1}^{n} a_k \phi_k \right\| \geqslant \sum_{i=1}^{n+1} \tilde{c}_i \left| l(x_i) - \sum_{k=1}^{n} a_k \phi_k(x_i) \right|$$
$$\geqslant \left| \sum_{i=1}^{n+1} \tilde{c}_i l(x_i) - \sum_{k=1}^{n} a_k \sum_{i=1}^{n+1} \tilde{c}_i \phi_k(x_i) \right| = \left| \sum_{i=1}^{n+1} \tilde{c}_i l(x_i) \right|$$
$$\geqslant \frac{1}{2n} \sum_{i=1}^{n+1} |\tilde{c}_i| = \frac{1}{2n}. \quad \text{Q.E.D.}$$

A particular n-dimensional subspace is $L(\Delta_n)$ and as a corollary of the Peano kernel theorem (cf. [16]), we have the following upper bound.

THEOREM 3.2. $\sup_{f \in A} \inf_{l \in L(\Delta_n)} \|f - l\| \leqslant 1/2(n-1)$.

Since $\frac{1}{2}(n-1)^{-1} = \frac{1}{2}n^{-1} + \frac{1}{2}(n(n-1))^{-1}$, we have that $L(\Delta_n)$ is essentially an optimal n-dimensional subspace of $C[0, 1]$ with respect to A.

4. Discrete Tchebycheff approximation. In view of the results of §3, we will concentrate on finding "good" mappings $M(\Delta)$: $C[0, 1] \to L(\Delta)$ for all sets of knots Δ: $0 = x_1 < x_2 < \cdots < x_n = 1$. By "good" we mean that (a) there exists a positive constant K such that

$$\|f - M(\Delta)f\| \leqslant K d(f, L(\Delta)), \text{ for all } f \in C[0, 1] \text{ and all } \Delta, \tag{4.1}$$

and (b) $M(\Delta)f$ is inexpensive to complete, i.e., the number of function evaluations and arithmetic operations needed to compute $M(\Delta)f$ is $O(n^p)$, p a positive integer, for all $f \in C[0, 1]$.

In this section, we describe and analyze an algorithm, which in the context of polynomial subspaces is due to de la Vallée Poussin (1919) (cf. [4]). If Ω is any subset of $[0, 1]$ and $f \in C[0, 1]$, we will use the notation $\|f\|_\Omega \equiv \max\{|f(x)| \mid x \in \Omega\}$. In particular, if $Y \equiv \{y_k \mid 0 \leqslant k \leqslant m\} \subset [0, 1]$ is a discrete point set, then we may consider the discrete Tchebycheff problem by finding $l_Y \in L(\Delta)$ which minimizes $\|f - l\|_Y \equiv \max_{y \in Y} |f(y) - l(y)|$ over all $l \in L(\Delta)$ and we define $M_Y(\Delta)f \equiv l_Y$. The problem of constructing l_Y is equivalent to a standard linear programming problem (cf. [15]), which may be solved by the simplex algorithm.

We now analyze this algorithm.

THEOREM 4.1. *If*

$$Y_i \equiv Y \cap [x_i, x_{i+1}] \neq 0, \quad |Y_i| \equiv \max_{x \in [x_i, x_{i+1}]} \min_{y \in Y_i} |x - y|,$$

and

$$|Y_i| < \tfrac{1}{2}(x_{i+1} - x_i), \quad 1 \leq i \leq n-1,$$

then

$$\|f - M_Y(\Delta)f\| \leq [2(1 - 2(x_{i+1} - x_i)^{-1}|Y_i|)^{-1} + 1]\, d(f, L(\Delta)) \quad \textit{for all } f \in C[0,1]. \tag{4.2}$$

To prove this result we use a basic inequality of A. A. Markov (1889) (cf. [11]) for polynomials. We state and prove Markov's inequality for the deceptively simple case of linear polynomials.

LEMMA. *If $l(x)$ is a linear polynomial on $[a, b]$, then*

$$\|Dl\|_{[a,b]} \leq 2(b-a)^{-1}\|l\|_{[a,b]}. \tag{4.3}$$

PROOF. We clearly have

$$|Dl(x)| = (b-a)^{-1}|l(b) - l(a)| \leq (b-a)^{-1}(|l(b)| + |l(a)|)$$
$$\leq 2(b-a)^{-1}\|l\|_{[a,b]}. \quad \text{Q.E.D.}$$

PROOF OF THEOREM 4.1. Let $t \in [x_i, x_{i+1}]$ be such that $|\tilde{l}(t) - l_Y(t)| = \|\tilde{l} - l_Y\|$ where $\tilde{l}$ is a best approximation to f in $L(\Delta)$. By the mean value theorem,

$$\begin{aligned} \|\tilde{l} - l_Y\|_{[x_i, x_{i+1}]} &= |\tilde{l}(t) - l_Y(t)| \\ &\leq \|\tilde{l} - l_Y\|_{Y_i} + |Y_i|\, \|D(\tilde{l} - l_Y)\|_{[x_i, x_{i+1}]}. \end{aligned} \tag{4.4}$$

Using Markov's inequality (4.3) to bound the right-hand side of (4.4), we have

$$\|\tilde{l} - l_Y\|_{[x_i, x_{i+1}]} \leq \|\tilde{l} - l_Y\|_{Y_i} + 2(x_{i+1} - x_i)^{-1}|Y_i|\, \|\tilde{l} - l_Y\|_{[x_i, x_{i+1}]} \tag{4.5}$$

and hence

$$\|\tilde{l} - l_Y\|_{[x_i, x_{i+1}]} \leq (1 - 2(x_{i+1} - x_i)^{-1}|Y_i|)^{-1}\|\tilde{l} - l_Y\|_{Y_i}. \tag{4.6}$$

Since $\|\tilde{l} - f\|_{Y_i} \leq \|\tilde{l} - f\|$ and $\|l_Y - f\|_{Y_i} \leq \|\tilde{l} - f\|_Y \leq \|\tilde{l} - f\|$, we have, from inequality (4.6),

$$\begin{aligned} \|\tilde{l} - l_Y\| &= \|\tilde{l} - l_Y\|_{[x_i, x_{i+1}]} \\ &\leq (1 - 2(x_{i+1} - x_i)^{-1}|Y_i|)^{-1} \cdot (\|\tilde{l} - f\|_{Y_i} + \|f - l_Y\|_{Y_i}) \\ &\leq (1 - 2(x_{i+1} - x_i)^{-1}|Y_i|)^{-1}(\|\tilde{l} - f\| + \|f - l_Y\|) \\ &\leq 2(1 - 2(x_{i+1} - x_i)^{-1}|Y_i|)^{-1}\|\tilde{l} - f\|. \end{aligned} \tag{4.7}$$

Combining (4.7) and the triangle inequality, we have

$$\|f - l_Y\| \leq \|f - \tilde{l}\,\| + \|\tilde{l} - l_Y\|$$
$$\leq [1 + 2(1 - 2(x_{i+1} - x_i)^{-1}|Y_i|)^{-1}]\ \|f - \tilde{l}\,\|,$$

which yields (4.2). Q.E.D.

Unfortunately we can show that the simplex algorithm for constructing l_Y will require $O(n^2)$ arithmetic operations (not counting function evaluations). As for function evaluations, there are two regimes to investigate. The first is where we can compute f at arbitrary points and we wish to economize, i.e., we want to minimize the number of functions evaluations. The second is where we are a priori given large quantities (relative to n) of data or approximate function evaluations— a situation typically arising in the analysis of experimental data which we wish to smooth and compress.

We will discuss the first regime in the remainder of this section. The second regime will be discussed in §5.

By Theorem 4.1, we need $|Y_i| < \frac{1}{2}(x_{i+1} - x_i)$ for all $1 \leq i \leq n-1$,which implies that we need at least two function evaluations in the interior of each element $[x_i, x_{i+1}]$. By symmetry, we minimize the coefficient of the right-hand side of (4.2) by evaluating the function f at the points $\frac{1}{4}(3x_i + x_{i+1})$ and $\frac{1}{4}(x_i + 3x_{i+1})$. With this choice of evaluation points for each element, $(x_{i+1} - x_i)^{-1}|Y_i| = \frac{1}{4}$ and using (4.2) we obtain the following result:

COROLLARY. *If* $Y \equiv \{y_k\}_{k=1}^{2n}$ *where* $y_{2i-1} \equiv \frac{1}{4}(3x_i + x_{i+1})$ *and* $y_{2i} \equiv \frac{1}{4}(x_i + 3x_{i+1})$, $1 \leq i \leq n$,

$$\|f - M_Y(\Delta)f\|_\infty \leq 5\, d(f, L(\Delta)). \tag{4.8}$$

This algorithm requires $2n$ evaluations of f.

It might be rather surprising that we can obtain a better result with essentially half of the function evaluations and no arithmetic operations! Suppose we evaluate f only at the knots $\{x_i\}_{i=1}^n$, i.e., $Y \equiv \Delta$. Then the preceding analysis does not quite hold since $|Y_i| = \frac{1}{2}(x_{i+1} - x_i)$. However, in this case, the discrete Tchebycheff problem is trivial to solve. In fact, its solution is the piecewise linear interpolate $I_{L(\Delta)}f$ of f. That is, if $x \in [x_i, x_{i+1}]$,

$$M_\Delta(\Delta)f(x) \equiv I_{L(\Delta)}f(x) \equiv (x_{i+1} - x_i)^{-1}[f(x_i)\,(x_{i+1} - x) + f(x_{i+1})\,(x - x_i)].$$

Fortunately, by a different technique, we can prove a version of (4.2) for this set of data.

THEOREM 4.2. *If* $f \in C[0, 1]$,

$$\|f - I_{L(\Delta)}f\| \leq 2d(f, L(\Delta)). \tag{4.9}$$

PROOF. If $f \in L(\Delta)$, the result is trivial. Otherwise, let $x \in [x_i, x_{i+1}]$ be such that $|f(x) - I_{L(\Delta)}f(x)| = \|f - I_{L(\Delta)}f\|$.

If $e(x) \equiv f(x) - I_{L(\Delta)}f(x)$, then clearly $e(x_i) = e(x_{i+1}) = 0$ and $\inf_{l \in L(\Delta)}\|f - l\| = \inf_{l \in L(\Delta)}\|e - l\|$. Thus, it suffices to show that, for all $l \in L(\Delta)$,

$$\|e - l\| \geqslant \|e - l\|_{[x_i, x_{i+1}]} \geqslant \tfrac{1}{2}\|e\|_{[x_i, x_{i+1}]} = \tfrac{1}{2}\|e\|.$$

If $|e(x) - l(x)| \geqslant \frac{1}{2}\|e\|_{[x_i, x_{i+1}]}$ for all $l \in L(\Delta)$, we are done.

Otherwise, $e(x)$ and $l(x)$ have the same sign and $|l(x)| \geqslant \frac{1}{2}\|e\|_{[x_i, x_{i+1}]}$. This implies that $|l(x_k)| \geqslant \frac{1}{2}\|e\|_{[x_i, x_{i+1}]}$ for $k =$ either i or $i + 1$.

Since $e(x_i) = e(x_{i+1}) = 0$, this implies that either

$$|e(x_i) - l(x_i)| \geqslant \tfrac{1}{2}\|e\|_{[x_i, x_{i+1}]} \quad \text{or} \quad |e(x_{i+1}) - l(x_{i+1})| \geqslant \tfrac{1}{2}\|e\|_{[x_i, x_{i+1}]}$$

Q.E.D.

This is the ideal situation for those problems in which we can evaluate a function arbitrarily. This interpolation algorithm has additional desirable features, such as linearity, which will be considered in detail in the next section.

5. Least squares algorithms. The problems in which we have a large quantity of data which we wish to smooth and compress have not been satisfactorily resolved as yet. Moreover, it is of further interest to have a projection algorithm, that is, one based on projection mappings. To be precise, a linear approximation algorithm is said to be a linear projection algorithm if and only if the associated mappings $M(s)$ are linear and are such that $M(s)y = y$ for all $y \in B(s)$, i.e., the mappings $M(s)$ are linear projectors.

Following ideas of Kantorovich, Lax, and deBoor (cf. [7]), we have the following equivalence result for linear projection algorithms:

THEOREM 5.1. *Let the sequence of subspaces* $\{B_n | 1 \leqslant n < \infty\}$ *be such that* $\lim_{n\to\infty}\inf_{b_n \in B_n}\|b - b_n\| = 0$ *for all* $b \in B$ *and the sequence of mappings* $\{M_n | 1 \leqslant n < \infty\}$ *be linear projectors which are consistent, i.e., there exists a dense subspace* D *of* B *such that* $\lim_{n\to\infty}\|M_n b - b\| = 0$ *for all* $b \in D$. *The following conditions are equivalent*:

(a) *There exists* $\epsilon > 0$ *such that* $\|M_n\| \leqslant \epsilon$ *for all* n.

(b) $\|b - M_n b\| \leqslant (1 + \epsilon)d(b, B_n)$, *for all* n *and* $b \in B$.

(c) $\|b - M_n b\| \to 0$ *as* $n \to \infty$ *for all* $b \in B$.

PROOF. (a) $\Rightarrow$ (b) For all $g \in B_n$,

$$\|b - M_n b\| \leqslant \|b - g\| + \|g - M_n b\| = \|b - g\| + \|M_n(g - b)\| \leqslant (1 + \epsilon)\|b - g\|.$$

(b) $\Rightarrow$ (c) is obvious.

(c) ⇒ (a) If $b \in B$, then $\{\|M_n b\| \mid n \geq 1\}$ is bounded. For otherwise there exists a sequence of elements $M_{n_1}b, \cdots, M_{n_j}b$ whose norms tend to infinity with contradicts convergence. Condition (a) follows from the uniform boundedness principle of functional analysis. Q.E.D.

This equivalence theorem is quite general. Let us consider some elementary applications.

EXAMPLES. (1) $B \equiv C[0, 1]$, $B_\Delta \equiv L(\Delta)$, and $M(\Delta) \equiv I_{L(\Delta)}$. Then $\|I_{L(\Delta)}\| = 1$ and hence $\|f - I_{L(\Delta)}f\| \leq 2d(f, L(\Delta))$, which is the result of Theorem 4.2.

(2) Let $B \equiv C(U)$, $\rho \equiv \Delta_x \times \Delta_y$ be the product grid on the unit square U and consider the tensor product space $B_\rho \equiv L(\rho) \equiv L(\Delta_x) \otimes L(\Delta_y)$. If $I_{L(\rho)}$ denotes the two-dimensional interpolation mapping into $L(\rho)$ (cf. [16]), then $\|I_{L(\rho)}\| = 1$ and hence $\|f - I_{L(\rho)}f\| \leq 2d(f, L(\rho))$.

(3) Let T be a triangulated polygon, $L(T)$ be the space of continuous, piecewise linear polynomials with respect to T, and $I_{L(T)}$ be the obvious mapping of $B \equiv C(T)$ into $B_T \equiv L(T)$. Then, once again $\|I_{L(T)}\| = 1$ and $\|f - I_{L(T)}f\| \leq 2d(f, L(T))$.

We now consider least squares algorithms. Given $f \in C[0, 1]$, we determine $\tilde{l} \in L(\Delta)$ which minimizes $\int_0^1 \int_0^1 (f(x) - l(x))^2\, dx \equiv \|f - l\|_2^2$ over $l \in L(\Delta)$; i.e., $\tilde{l} \equiv P(\Delta)f$ is the orthogonal projection of f onto $L(\Delta)$. Expressing $\tilde{l}(x) = \Sigma_{i=1}^n \tilde{\beta}_i l_i(x)$, where $l_i(x)$ is the unique element in $L(\Delta)$ defined by

$$l_i(x_j) = \delta_{ij} \equiv \begin{cases} 1 & \text{if } i = j, \\ 0 & \text{otherwise,} \end{cases}$$

the minimization problem is equivalent to solving

$$A\tilde{\beta} = \mathbf{k}, \tag{5.1}$$

where A is the $n \times n$ matrix given by

$$A \equiv \begin{bmatrix} x_2/3 & \cdot & & & & \\ \cdot & \cdot & \cdot & & & 0 \\ & \cdot & \cdot & \cdot & & \\ & (x_i - x_{i-1})/6 & (x_{i+1} - x_{i-1})/3 & (x_{i+1} - x_i)/6 & & \\ & & \cdot & \cdot & \cdot & \\ 0 & & & \cdot & \cdot & (1 - x_{n-1})/3 \end{bmatrix} \tag{5.2}$$

and

$$\mathbf{k} \equiv \left[\int_0^1 f(x) l_i(x)\, dx\right]. \tag{5.3}$$

For the case of a uniform mesh with $h \equiv x_{i+1} - x_i$, $1 \leq i \leq n-1$,

$$A = \frac{h}{6}\begin{bmatrix} 2 & 1 & & 0 \\ 1 & 4 & \ddots & \\ & \ddots & 4 & 1 \\ 0 & & 1 & 2 \end{bmatrix}.$$

Using the elementary fact that $\|\Sigma_{i=1}^n \beta_i l_i\| = \|\beta\|_\infty \equiv \max_{1 \leq i \leq n} |\beta_i|$ for all $\beta \in R^n$, we can obtain the following result for the mappings P_Δ:

THEOREM 5.2. *For all* $f \in C[0, 1]$, $\|f - P_\Delta f\| \leq 4d(f, L(\Delta))$.

PROOF. If $D \equiv [d_{ij}]$ is the diagonal matrix with $d_{11} \equiv 3x_2^{-1}$, $d_{ii} \equiv 3(x_{i+1} - x_{i-1})^{-1}$, $2 \leq i \leq n-1$, and $d_{nn} \equiv 3(1 - x_{n-1})^{-1}$, i.e., D^{-1} is the diagonal of A, then $A\hat{\beta}' = \mathbf{k}$ implies

$$DA\hat{\beta}' = D\mathbf{k}, \tag{5.4}$$

where $DA \equiv I + M$, $\|M\|_\infty \equiv \max_{1 \leq i \leq n} \Sigma_{j=1}^n |m_{ij}| = ½$, and $\|Dk\|_\infty \leq 3\|f\|/2$.

It follows that $\|(DA)^{-1}\|_\infty = \|(I + M)^{-1}\|_\infty \leq 2$ (cf. [1]), and hence that $\|\hat{\beta}'\|_\infty \leq 3\|f\|$. The result now follows from Theorem 5.1. Q.E.D.

What about computing (5.4) and its solution? If we can compute the integrals needed to evaluate $\mathbf{k}$, then we need only $O(n)$ arithmetic operations to compute the linear system (5.4). Moreover, we can solve the tridiagonal linear system in $O(n)$ arithmetic operations by means of Gaussian elimination for tridiagonal matrices (cf. [1]).

From the viewpoint of round-off error analysis, it is important to know about the conditioning of DA. In this case, it turns out that the condition number of DA, i.e., $\|(DA)^{-1}\|_\infty \|DA\|_\infty$, is uniformly bounded (independent of Δ) by $2 \cdot 3 = 6$. Thus, we have an ideal situation.

To handle discrete data, we suggest an algorithm of Patent (cf. [12]). We note that the data occurs only in the right-hand side of the linear system (5.1). Thus, if the data is at Y, we choose $\Delta \subset Y$, interpolate f at Y by $I_{L(Y)}f$, replace f in the right-hand side of (5.1), and compute the resulting integrals whose integrands are piecewise quadratics by either symbolic methods or Gaussian quadrature with two nodes in each element defined by Y. That is, we solve the linear system

$$A\hat{\beta} = \hat{\mathbf{k}} \equiv \left[\int_0^1 I_{L(Y)}f(x) l_i(x)\, dx\right],$$

and let $\hat{M}(\Delta)f \equiv \Sigma_{i=1}^n \hat{\beta}_i l_i(x)$. We can prove the following error bound for this procedure.

THEOREM 5.3. *If* $\Delta \subset Y$, *then, for all* $f \in C[0, 1]$,

(5.5) $$\|f - \hat{M}(\Delta)f\| \leqslant 4d(f, L(\Delta)).$$

PROOF. By Thoerem 5.1, it suffices to show that $\|\hat{M}_\Delta\| \leqslant 3$; since $\Delta \subset Y, L(\Delta) \subset L(Y)$ and $\hat{M}(\Delta) = P_\Delta I_{L(Y)}$.

Hence, $\|\hat{M}(\Delta)\| \leqslant \|P_\Delta\| \, \|I_{L(Y)}\| \leqslant 3 \cdot 1 = 3$. Q.E.D.

6. A two-dimensional extension. In this section, we consider the problem of approximating a function $f(x, y)$ of two variables on the unit sphere square U by means of bilinear functions in $L(\rho)$, where $\rho \equiv \Delta_x \times \Delta_y$ is a product rectangular mesh. Our least squares algorithm seeks $P_\rho f \equiv \hat{l}(x, y) \in L(\rho)$ which minimizes

(6.1) $$\int_0^1 \int_0^1 (f(x, y) - l(x, y))^2 \, dx \, dy$$

over $l \in L(\rho)$. If $\hat{l}(x, y) = \Sigma_{i=1}^n \Sigma_{j=1}^n \hat{\beta}_{ij} l_i(x) l_j(y)$, then it is easy to show that $\hat{\beta} \equiv \hat{\beta}_{n(i-1)+j} \equiv \hat{\beta}_{ij}$, $1 \leqslant i, j \leqslant n$, is the unique solution of the linear system

(6.2) $$A_x \otimes A_y \hat{\beta} = \mathbf{k},$$

where A_x and A_y are the one-dimensional least squares matrices with respect to $L(\Delta_x)$ and $L(\Delta_y)$ respectively and $A_x \otimes A_y$ is the Kronecker product, i.e.,

$$[a_{ij}] \otimes [b_{ij}] \equiv \begin{bmatrix} a_{11}[b_{ij}] & \cdots & a_{1n}[b_{ij}] \\ \vdots & \ddots & \vdots \\ a_{n1}[b_{ij}] & \cdots & a_{nn}[b_{ij}] \end{bmatrix}$$

Note that we have used the "natural ordering along vertical lines of ρ" for the solution vector $\hat{\beta}$. The matrix $A_x \otimes A_y$ is sparse. In fact, it has only nine nonzero diagonals and a band width of $n + 2$. For the special case of $n = 2$, its associated graph is

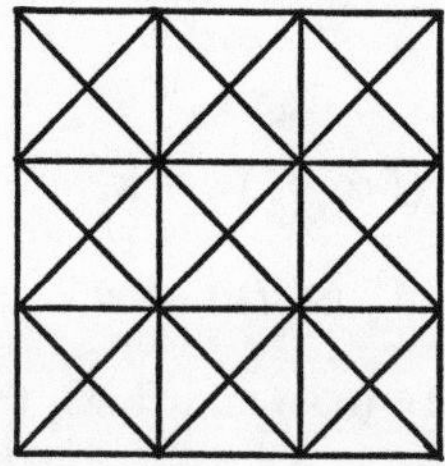

Of course, we would like to prove an error bound which is a two-dimensional analogue of Theorem 5.2. We may do this with the aid of some elementary results about Kronecker products.

THEOREM 6.1. *For all* $f \in C(U)$,

$$\|f - P_\rho f\| \leqslant 10d(f, L(\rho)). \tag{6.3}$$

PROOF. If D_x and D_y are the diagonal matrices defined in the proof of Theorem 5.2, then

$$(D_xA_x \otimes D_yA_y)\tilde{\beta} = (D_x \otimes D_y)(A_x \otimes A_y)\tilde{\beta} = (D_x \otimes D_y)\mathbf{k},$$

where we have used the fact that the product of tensor products is the tensor product of the products (cf. [10]). As in the proof of Theorem 5.2, $D_xA_x \equiv I + M_x$ and $D_yA_y \equiv I + M_y$ where $\|M\|_\infty = 5/4$. Hence, $D_xA_x \otimes D_yA_y$ is *not* diagonally dominant. However, the inverse of a tensor product is the tensor product of the inverses and the ∞-norm of a tensor product is less than or equal to the product of the ∞-norms. Thus

$$\|(D_xA_x \otimes D_yA_y)^{-1}\|_\infty \leqslant \|(D_xA_x)^{-1}\|_\infty \|(D_yA_y)^{-1}\|_\infty \leqslant 2 \cdot 2 = 4.$$

Moreover, $\|\mathbf{k}\|_\infty \leqslant 9\|f\|/4$. Hence, $\|P_\rho f\| = \|\beta\|_\infty \leqslant 4 \cdot 9\|f\|/4 = 9\|f\|$, and the result follows from Theorem 5.1. Q.E.D.

If the data is given as a discrete function on a rectangular grid Y including ρ as a subset, we may prove a two-dimensional analogue of Theorem 5.3. The details are left to the reader.

An interesting and important issue is the choice of algorithms for solving linear systems of the form (6.2). If we use band or profile Gaussian elimination, we need $O(n^3)$ storage locations and $O(n^4)$ arithmetic operations (cf. [6]). If we use sparse matrix techniques, then the best we can do is $O(n^2 \ln n)$ storage locations and $O(n^3)$ arithmetic operations (cf. [5] and [6]). These latter results hold for J. A. George's "nested ordering" of the unknowns.

However, using the special structure of the equations, we can achieve an "alternating direction" direct method which requires $O(n^2)$ storage locations and $O(n^2)$ arithmetic operations. To start we observe that it suffices to solve the coupled systems

$$(I \otimes A_y)\mathbf{w} = \mathbf{k}, \tag{6.4}$$

$$(A_x \otimes I)\tilde{\beta} = \mathbf{w}. \tag{6.5}$$

In fact, if $\mathbf{w}$ and $\tilde{\beta}$ satisfy (6.4) and (6.5), then $(I \otimes A_y)(A_x \otimes I)\tilde{\beta} = (I \otimes A_y)\mathbf{w} = \mathbf{k}$. But

$$I \otimes A_y \equiv \begin{bmatrix} A_y & & \mathbf{0} \\ & \ddots & \\ \mathbf{0} & & A_y \end{bmatrix}$$

is an $n \times n$ block diagonal matrix with $n \times n$ blocks. If we partition $\mathbf{w}$ and $\mathbf{k}$ into the corresponding n-block vectors, we have

$$A_y \mathbf{w}_l = \mathbf{k}_l, \quad 1 \leq l \leq n. \tag{6.6}$$

Moreover, each system in (6.6) can be solved with $O(n)$ storage locations and $O(n)$ arithmetic operations. Since there are n such systems, we need a total of $O(n^2)$ storage locations and $O(n^2)$ arithmetic operations to compute $\mathbf{w}$.

To solve (6.5) efficiently, we define $\hat{\beta}$ and $\hat{\mathbf{w}}$ by reordering the components of $\tilde{\beta}$ and $\mathbf{w}$ to correspond to the "natural ordering along *horizontal* lines of ρ," e.g., $\hat{\beta}_{n(j-1)+i} \equiv \tilde{\beta}_{ij}$, $1 \leq i, j \leq n$. Then, (6.5) may be rewritten as

$$(I \otimes A_x)\hat{\beta} = \hat{\mathbf{w}}. \tag{6.7}$$

This system may be solved the same way we solved (6.6) with $O(n^2)$ storage locations and $O(n^2)$ arithmetic operations, and $\tilde{\beta}_{ij}$, $1 \leq i, j \leq n$, may be reconstructed from $\hat{\beta}$.

A moral of this analysis is that a sparse linear system with a special structure may often be solved more efficiently by means of a special algorithm than by general sparse matrix algorithms.

Acknowledgement. The author is grateful to Professor S. C. Eisenstat for many helpful discussions regarding the content of this paper.

References

1. E. K. Blum, *Numerical analysis and computation: Theory and practice*, Addison-Wesley, Reading, Mass., 1972.

2. C. de Boor, *On the convergence of odd-degree spline interpolation*, J. Approximation Theory 1 (1968), 452–463. MR 38 #6273.

3. ———, *On uniform approximation by splines*, J. Approximation Theory 1 (1968), 219–235. MR 39 #1866.

4. C. de la Vallée Poussin, *Leçons sur l'approximation des fonctions d'une variable réelle*, Gauthier-Villars, Paris, 1919.

5. J. A. George, *Nested disection of a regular finite element mesh*, SIAM J. Numer. Anal. 10 (1973), 345–363.

6. A. J. Hoffman, M. S. Martin and D. J. Rose, *Complexity bounds for regular finite difference and finite element grids*, SIAM J. Numer. Anal. 10 (1973), 364–369.

7. L. V. Kantorovič, *Functional analysis and applied mathematics*, Uspehi Mat. Nauk 3 (1948), no. 6 (28), 89–185; English transl., Nat. Bur. Standards Rep., no. 1509, U. S. Dept. of Commerce, Nat. Bur. Standards, Washington, D.C., 1952. MR 10, 380; 14, 766.

8. A. N. Kolmogorov, *Über die beste Annäherung von Funktionen einer gegebenen Funktionenklasse*, Ann. of Math. 37 (1936), 107–111.

9. G. G. Lorentz, *Approximation of functions*, Holt, Rinehart, and Winston, New York, 1966. MR 35 #4642.

10. M. Marcus and H. Minc, *A survey of matrix theory and matrix inequalities,* Allyn and Bacon, Boston, Mass., 1964. MR 29 #112.

11. A. A. Markov, *On a problem of D. I. Mendeleev,* St. Petersburg Izv. Akad. Nauk 62 (1889), 1–24.

12. P. D. Patent, *Least square polynomial spline approximation,* Doctoral Thesis, California Institute of Technology, 1972.

13. J. R. Rice, *On the degree of convergence of nonlinear spline approximation,* Approximations with Special Emphasis on Spline Functions (Proc. Sympos. Univ. of Wisconsin, Madison, Wis., 1969), Academic Press, New York, 1970. MR 42 #2226.

14. ———, *On the computational complexity of approximation operators,* Austin meeting on Approximation Theory, January 1973.

15. M. H. Schultz, *Discrete Tchebycheff approximation for multivariate splines,* J. Comput. System Sci. 6 (1972), 298–304. MR 46 #5898.

16. ———, *Spline analysis,* Prentice-Hall, Englewood Cliffs, N. J., 1973.

17. V. M. Tihomirov, *Diameters of sets in functional spaces and the theory of best approximations,* Uspehi Mat. Nauk 15 (1960), no. 3 (93), 81–120 = Russian Math. Surveys 15 (1960), no. 3, 75–111. MR 22 #8268.

18. J. F. Traub, *Computational complexity of iterative processes,* SIAM J. Comput. 1 (1972), 167–179. MR 47 #2860.

YALE UNIVERSITY

SIAM-AMS Proceedings
Volume 7
1974

Computational Complexity of One-Point and Multi-Point Iteration*

H. T. Kung and J. F. Traub

Abstract. Let φ be an iteration for approximating the solution of a problem f. We define a new efficiency measure $e(\varphi, f)$. For a given problem f, we define the optimal efficiency $E(f)$ and establish lower and upper bounds for $E(f)$ with respect to different families of iterations. We conjecture an upper bound on $E(f)$ for any iteration without memory.

1. Introduction. Let φ be an iteration for approximating the solution of a problem f. We define a new efficiency measure $e(\varphi, f)$. The efficiency measure gives us a methodology for comparing iterations as well as permitting us to derive theoretical limits on iteration efficiency.

For a given problem f, we define the optimal efficiency $E(f)$ over all φ belonging to a family Φ. We establish lower and upper bounds for $E(f)$ with respect to different families of iterations. We conjecture an upper bound on $E(f)$ for any iteration without memory.

We summarize the results of this paper. Basic concepts are given in §2 and our efficiency measure is defined in §3. In the next two sections we establish lower and upper bounds on the optimal efficiency for solving a problem with respect to important families of algorithms. A conjecture on optimal efficiency is stated in §6 and a small numerical example is given in the last section.

2. Basic concepts. We work over the field of real numbers. Let $\sigma(x)$ be a function and λ_σ be a procedure which computes the value of $\sigma(x)$ for any given x. (We write λ for λ_σ if there is no ambiguity.) Let α be any number. We say $\Sigma = (\sigma, \lambda)$ is an *algorithm* for approximating α if the sequence $\{x_i\}$, generated by $x_{i+1} = \sigma(x_i)$, converges to α whenever x_0 is chosen near α,

---

AMS (MOS) subject classifications (1970). Primary 68A20, 62A10, 65H05.

*This research was supported in part by the National Science Foundation under grant GJ-32111 and the Office of Naval Research under contract N00014-67-A-0314-0010, NR 044-422.

and if $\sigma(x_i)$ is computed by the procedure λ for all i. $\Sigma = (\sigma, \lambda)$ has *order of convergence* $p(\sigma)$ if

$$\lim_{x \to \alpha} \frac{\sigma(x) - \alpha}{(x - \alpha)^{p(\sigma)}}$$

exists and is nonzero. We measure the goodness of the algorithm $\Sigma = (\sigma, \lambda)$ by $p(\sigma)$ and define the efficiency of the algorithm $\Sigma = (\sigma, \lambda)$ to be

$$e(\Sigma) = \frac{\log p(\sigma)}{c(\lambda)}, \tag{2.1}$$

where $c(\lambda)$ is the cost of performing the procedure λ. In this paper we consider only superlinear convergent algorithms, that is, $p(\sigma) > 1$. All logarithms are to base 2.

For any fixed positive integer n, consider the algorithm $\Sigma_n = (\sigma_n, \lambda_n)$ where $\sigma_n = \sigma \circ \sigma \circ \cdots \circ \sigma$ (σ occurs n times and $\sigma \circ \sigma$ denotes composition) and λ_n is the procedure which computes $\sigma_n(x)$ by

$$y_0 = x,$$

$$y_{i+1} = \sigma(y_i), \qquad i = 0, \cdots, n-1,$$

$$\sigma_n(x) = y_n,$$

with $\sigma(y_i)$ being computed by λ for all i. One can easily check that $p(\sigma_n) = p^n(\sigma)$ and $c(\lambda_n) = nc(\lambda)$. Note that

$$\frac{\log p(\sigma)}{c(\lambda)} = \frac{\log p^n(\sigma)}{nc(\lambda)}.$$

Therefore, $e(\Sigma) = e(\Sigma_n)$ for any n. This invariance is clearly desirable for any useful efficiency measure, since Σ_n is just the algorithm which repeats Σ n times and hence Σ and Σ_n must have the same efficiency. Gentleman [1] shows that if any efficiency measure satisfies this invariance property then it must be of the form (2.1) or a strictly increasing function of that form. Hence (2.1) is essentially the unique way to define an efficiency measure. Furthermore, Traub [8, Equation C-11] shows that if the efficiency measure has the form (2.1) then efficiency is inversely proportional to the total cost of approximating α by the algorithm. More specifically, let Σ^1, Σ^2 be two algorithms for approximating α and let $k(\Sigma^1), k(\Sigma^2)$ be the total costs for generating two sequences which start with the same initial approximation and terminate when some fixed number of correct digits of α have been calculated. Then

$$k(\Sigma^1)/k(\Sigma^2) \sim e(\Sigma^2)/e(\Sigma^1). \tag{2.2}$$

Therefore, it is desirable to have algorithms with high efficiency. *An algorithm is called optimal in a certain class of algorithms if it has the highest efficiency among all algorithms in that class.*

We now consider how to define the cost $c(\lambda)$. Paterson [7] defines $c(\lambda)$ as the number of multiplications or divisions, except by constants, needed to perform the procedure λ. We call the associated efficiency the multiplicative efficiency. Kung [3] shows that unity is the sharp upper bound on the multiplicative efficiency, and Kung [4] uses the multiplicative efficiency to investigate the computational complexity of algebraic numbers. In this paper, *we define $c(\lambda)$ to be the number of arithmetic operations needed to perform the procedure λ.*

3. Efficiency measure for iteration. In the previous section we have defined the efficiency of an algorithm for approximating a number α. More specifically, we now study the efficiency of an algorithm for approximating a simple zero α_f of a function $f \in D$, where D is the set of analytic functions f which have simple zeros α_f. We consider algorithms $\Sigma = (\sigma, \lambda)$ where $\sigma = \varphi(f)$, φ is a one-point or multi-point iteration and $f \in D$. (See Kung and Traub [5].) If φ is a k-point iteration, $k = 1, 2, \cdots$, then φ has the following property:

There exist nonnegative integers $d_0, \cdots, d_{k-1}$ and functions

$$u_{j+1}(y_0; y_1^0, \cdots, y_{d_0+1}^0; \cdots; y_1^j, \cdots, y_{d_j+1}^j)$$

of $1 + \Sigma_{i=0}^{j}(d_i + 1)$ variables for $j = -1, \cdots, k-1$ such that, for all $f \in D$, if

$$(3.1)\quad \begin{cases} z_0(x) = x, \quad x \text{ belongs to the domain of } \varphi(f), \\ y_{i+1}^j(x) = f^{(i)}(z_j(x)), \\ z_{j+1}(x) = u_{j+1}(x; y_1^0(x), \cdots, y_{d_0+1}^0(x); \cdots; y_1^j(x), \cdots, y_{d_j+1}^j(x)), \\ \qquad\qquad\qquad \text{for } j = 0, \cdots, k-1, i = 0, \cdots, d_j, \end{cases}$$

then

$$(3.2)\qquad \varphi(f)(x) = z_k(x).$$

In this paper we assume that

(3.3) all u_j are rational functions;

(3.4) if f is transcendental, we use a rational subroutine to approximate $f^{(i)}, i \geq 0$, whenever $f^{(i)}$ is transcendental;

(3.5) all $f^{(i)}(z_j(x))$ are algebraically independent.

Assumption (3.5) means that we are not allowed to use any special property of f. In other words, we consider "general" f.

Recall that λ $(= \lambda_{\varphi(f)})$ is a procedure which computes the value of $\varphi(f)(x)$ for any x. Because of (3.5), λ must compute $\varphi(f)(x)$ according to (3.1) and (3.2). Let $a_j(\lambda), j = 1, \cdots, k$, denote the number of arithmetic operations needed to compute $u_j(y_0; y_1^0, \cdots, y_{d_{j-1}+1}^{j-1})$ for given $(y_0; y_1^0, \cdots, y_{d_{j-1}+1}^{j-1})$ by the procedure λ. Moreover, if $f^{(i)}$ is rational, let $c(f^{(i)})$ denote the number of arithmetic operations for one evaluation of $f^{(i)}$; otherwise let $c(f^{(i)})$ denote the number of arithmetic operations used in the rational subroutine which approximates $f^{(i)}$. Then the total number of arithmetic operations needed to perform the procedure λ is

$$c(\lambda) = \sum_{i \geqslant 0} v_i(\varphi) c(f^{(i)}) + \sum_{i=1}^{k} a_i(\lambda)$$

where $v_i(\varphi)$ is the number of evaluations of $f^{(i)}$ required by φ.

If $p(\varphi)$ is the order of convergence of the iteration φ, then by definition (2.1) the efficiency of the algorithm $(\varphi(f), \lambda)$ is

$$e(\varphi(f), \lambda) = \frac{\log p(\varphi)}{c(\lambda)} = \frac{\log p(\varphi)}{\sum_{i \geqslant 0} v_i(\varphi) c(f^{(i)}) + \sum_{i=1}^{k} a_i(\lambda)}.$$

We define $e(\varphi, f)$, the *efficiency of the iteration* φ *with respect to the problem* f, by

$$e(\varphi, f) = \sup_{\lambda} e(\varphi(f), \lambda).$$

Let

$$a(\varphi) = \min_{\lambda} \sum_{i=1}^{k} a_i(\lambda).$$

Then

$$e(\varphi, f) = \frac{\log p(\varphi)}{\sum_{i \geqslant 0} v_i(\varphi) c(f^{(i)}) + a(\varphi)}. \tag{3.6}$$

This is the basic efficiency measure used in this paper.

Define $\sum_{i \geqslant 0} v_i(\varphi) c(f^{(i)})$ to be the *evaluation cost* of φ with respect to f and define $a(\varphi)$ to be the *combinatory cost* of φ. The total cost, which appears in the denominator of (3.6), is the sum of these two costs.

Let

$$c_f = \min_{i \geqslant 0} c(f^{(i)}). \tag{3.7}$$

In this paper, we refer to c_f *as the problem complexity.* Let

$$(3.8) \qquad v(\varphi) = \sum_{i \geq 0} v_i(\varphi).$$

Then, by (3.6),

$$(3.9) \qquad e(\varphi, f) \leq \frac{\log p(\varphi)}{v(\varphi)c_f + a(\varphi)}.$$

This gives an upper bound on $e(\varphi, f)$.

The efficiency measure defined by (3.6) is the first one to include both evaluation and combinatory costs. Ostrowski [6, Chapter 3] defines efficiency as $p(\varphi)^{1/v(\varphi)}$ where $v(\varphi)$ is defined by (3.8). This amounts to neglecting $a(\varphi)$ and taking $c(f^{(i)})$ to be unity for all i in (3.6). Our efficiency measure, defined by (3.6), does not take into account rounding errors or truncation errors caused by rational approximations for transcendental $f^{(i)}, i \geq 0$.

The following two examples illustrate the definitions.

EXAMPLE 3.1 (NEWTON-RAPHSON ITERATION).

$$\varphi(f)(x) = x - f(x)/f'(x).$$

This is a one-point iteration with $p(\varphi) = 2, v_0(\varphi) = v_1(\varphi) = 1$, and $a(\varphi) = 2$. Hence

$$e(\varphi, f) = \frac{1}{c(f) + c(f') + 2}, \qquad e(\varphi, f) \leq \frac{1}{2c_f + 2}.$$

EXAMPLE 3.2.

$$z_0 = x,$$

$$z_1 = z_0 - f(z_0)/f'(z_0),$$

$$\varphi(f)(x) = z_1 - \frac{f(z_1)f(z_0)}{[f(z_1) - f(z_0)]^2} \cdot \frac{f(z_0)}{f'(z_0)}.$$

This is a two-point iteration with $p(\varphi) = 4, v_0(\varphi) = 2, v_1(\varphi) = 1$ and $a(\varphi) = 8$. (See Kung and Traub [5, §5].) Hence

$$e(\varphi, f) = \frac{2}{2c(f) + c(f') + 8}, \qquad e(\varphi, f) \leq \frac{2}{3c_f + 8}.$$

It is natural to ask for a given problem f what is the optimal value of $e(\varphi, f)$ for all φ belonging to some family Φ. Define

$$E_n(\Phi, f) = \sup_{\varphi \in \Phi} \{e(\varphi, f) \mid v(\varphi) = n\}.$$

Thus $E_n(\Phi, f)$ is the optimal efficiency over all $\varphi \in \Phi$ which use n evaluations. Define

$$E(\Phi, f) = \sup \{E_n(\Phi, f) | n = 1, 2, \cdots\}.$$

Thus $E(\Phi, f)$ *is the optimal efficiency for all* $\varphi \in \Phi$. *We will establish lower and upper bounds for* $E_n(\Phi, f)$ *and* $E(\Phi, f)$ *with respect to different families of iterations.* When there is no ambiguity, we write $E_n(\Phi, f)$ and $E(\Phi, f)$ as $E_n(f)$ and $E(f)$, respectively. Since in practice we are more concerned with efficiency for problems f with higher complexity, we are particularly interested in the asymptotic behavior of these bounds as $c_f \to \infty$.

4. Theorems on efficiency of one-point iteration. We consider first the family of one-point iterations $\{\gamma_n\}$. (See Kung and Traub [5, §3].) The important properties of $\{\gamma_n\}$ from our point of view are summarized in the following theorem proven by Traub [8, §5.1].

THEOREM 4.1. 1. $v_i(\gamma_n) = 1, i = 0, \cdots, n-1, v_i(\gamma_n) = 0, i > n-1$. *Hence* $v(\gamma_n) = n$.

2. $p(\gamma_n) = n$.

We now turn to an upper bound for $a(\gamma_n)$. Suppose that we have already obtained $f^{(i)}(x), i = 0, \cdots, n-1$, and we want to use them to form $\gamma_n(f)(x)$. This amounts to calculating the first $n-1$ derivatives of f^{-1} (the inverse function) at $f(x)$. This can be done in $O(n^3)$ arithmetic operations by the power series reversion technique reported in Knuth [2, §4.7]. However if one uses the fast Fourier transform for polynomial multiplication then the power series reversion can be done in $O(n^2 \log n)$ arithmetic operations, and this implies that

$$a(\gamma_n) \leqslant \rho n^2 \log n \tag{4.1}$$

for some positive constant ρ. Then, by (4.1) and Theorem 4.1,

$$e(\gamma_n, f) \geqslant \frac{\log n}{\sum_{i=0}^{n-1} c(f^{(i)}) + \rho n^2 \log n}. \tag{4.2}$$

For n small, $a(\gamma_n)$ can be calculated by inspection. For instance, since

$$\gamma_3(f)(x) = x - \frac{f(x)}{f'(x)} - \frac{f''(x)}{2f'(x)} \left[\frac{f(x)}{f'(x)}\right]^2,$$

one can easily observe that $a(\gamma_3) = 7$. Hence

$$e(\gamma_3, f) = \frac{\log 3}{c(f) + c(f') + c(f'') + 7}. \tag{4.3}$$

Let φ be any one-point iteration, with $v(\varphi) = n$, which satisfies a mild smoothness condition. Then, by Traub [8, §5.4], Kung and Traub [5, Theorem 6.1], $v_i(\varphi) \geq 1, i = 0, \cdots, p(\varphi) - 1$, and hence $p(\varphi) \leq n$. Clearly, $a(\varphi) \geq n - 1$. Therefore, from (3.9),

$$e(\varphi, f) \leq \frac{\log n}{nc_f + n - 1} \equiv h(n). \tag{4.4}$$

It is straightforward to verify that

$$h(n) \leq \frac{\log 3}{3c_f + 2}, \quad \text{for } c_f > 4. \tag{4.5}$$

From (4.2), (4.3), (4.4) and (4.5) we have

THEOREM 4.2. *For the family* Φ *of one-point iterations,*

$$\frac{\log n}{\Sigma_{i=0}^{n-1} c(f^{(i)}) + \rho n^2 \log n} \leq E_n(f) \leq \frac{\log n}{nc_f + n - 1}, \quad \text{for a constant } \rho > 0, \forall n, \tag{4.6}$$

$$\frac{\log 3}{c(f) + c(f') + c(f'') + 7} \leq E(f) \leq \frac{\log 3}{3c_f + 2}, \quad \text{for } c_f > 4. \tag{4.7}$$

REMARK 4.1. 1. In (4.6) both lower and upper bounds for $E_n(f)$ are tight for f such that $c(f^{(i)}) \sim c_f, i < n$, and c_f is large, since lower bound/upper bound $\longrightarrow 1$ as $c_f \longrightarrow \infty$.

2. For f such that $c(f) \sim c(f') \sim c(f'') \sim c_f$, and c_f is large, both lower and upper bounds for $E(f)$ in (4.7) are tight, since lower bound/upper bound $\longrightarrow 1$ as $c_f \longrightarrow \infty$. In this case, by (4.3), γ_3 is close to optimal among all one-point iterations.

5. Theorems on efficiency of multi-point iteration. We consider first the family of iterations $\{\Psi_n\}$ defined by Kung and Traub [5, §4]. The important properties of $\{\Psi_n\}$ from our point of view are summarized in

THEOREM 5.1. 1. $v_0(\Psi_n) = n$. $v_i(\Psi_n) = 0, i > 0$. *Hence* $v(\Psi_n) = n$.
2. $p(\Psi_n) = 2^{n-1}$.

Kung and Traub [5, Appendix] give a procedure λ for computing $\Psi_n(f)(x)$. It can be shown that $\Sigma_{j=1}^n a_j(\lambda) = 3n^2/2 + 3n/2 - 7$. Hence $a(\Psi_n) \leq 3n^2/2 + 3n/2 - 7$. More generally, we assume that

$$a(\Psi_n) \leq r(n), \tag{5.1}$$

where $r(n) = r_2 n^2 + r_1 n + r_0, r_2 > 0$.

Then, by (5.1) and Theorem 5.1,

$$e(\Psi_n, f) \geq \frac{n-1}{nc(f) + r(n)}. \tag{5.2}$$

We choose n so as to maximize the right-hand side of (5.2). The maximum is achieved when $n = t$ where

$$t = 1 + (c(f)/r_2 + \delta)^{1/2}, \qquad \delta = (r_0 + r_1 + r_2)/r_2.$$

Let

$$M = \text{round } (t). \tag{5.3}$$

Then from (5.2) we can easily prove

THEOREM 5.2. *There exists a constant* $\zeta < 0$ *such that if* $M = M(f)$ *is chosen by* (5.3) *then*

$$e(\Psi_M, f) \geq \frac{1}{c(f)}\left(1 + \frac{\zeta}{(c(f))^{1/2}}\right), \quad \textit{for } c(f) \textit{ large.}$$

From (5.2) and Theorem 5.2, we have

COROLLARY 5.1. *For the family* Φ *of one-point or multi-point iterations,*

$$E_n(f) \geq \frac{n-1}{nc(f) + r(n)}, \quad \textit{where } r(n) = r_2 n^2 + r_1 n + r_0,\ r_2 > 0; \textit{ and}$$

$$E(f) \geq \frac{1}{c(f)}\left[1 + \frac{\zeta}{(c(f))^{1/2}}\right], \quad \textit{for a constant } \zeta < 0, \textit{ for } c(f) \textit{ large.}$$

We turn to the family of iterations $\{\omega_n\}$ defined in Kung and Traub [5, §5]. The important properties of $\{\omega_n\}$ from our point of view are summarized in

THEOREM 5.3. 1. $v_0(\omega_n) = n - 1, v_1(\omega_n) = 1, v_i(\omega_n) = 0, i > 1$. *Hence* $v(\omega_n) = n$.

2. $p(\omega_n) = 2^{n-1}$.

Kung and Traub [5, Appendix] give a procedure λ for computing $\omega_n(f)(x)$. It can be shown that

$$\sum_{j=1}^{n} a_j(\lambda) = \frac{3}{2}n^2 + \frac{3}{2}n - 4.$$

Hence $a(\omega_n) \leq 3n^2/2 + 3n/2 - 4$. More generally, we assume that

$$a(\omega_n) \leq s(n) \tag{5.4}$$

where $s(n) = s_2 n^2 + s_1 n + s_0, s_2 > 0$. Then, by (5.4) and Theorem 5.3,

$$e(\omega_n, f) \geqslant \frac{n-1}{(n-1)c(f) + c(f') + s(n)}. \tag{5.5}$$

We choose n so as to maximize the right-hand side of (5.5). Then the maximum is achieved when $n = u$, where

$$u = 1 + \left(\frac{c(f')}{s_2} + \epsilon\right)^{1/2}, \qquad \epsilon = \frac{s_0 + s_1 + s_2}{r_2}.$$

Let

$$N = \text{round}\,(u). \tag{5.6}$$

Then from (5.5) we can easily prove

THEOREM 5.4. *There exists a constant* $\eta > 0$ *such that if* $N = N(f)$ *is chosen by* (5.6) *then*

$$e(\omega_N, f) \geqslant \frac{1}{c(f) + \eta(c(f'))^{1/2}}, \quad \textit{for } c(f') \textit{ large.}$$

From (5.5) and Theorem 5.4, we have

COROLLARY 5.2. *For the family* Φ *of one-point or multi-point iterations,*

$$E_n(f) \geqslant \frac{n-1}{(n-1)c(f) + c(f') + s(n)}, \quad \textit{where } s(n) = s_2 n^2 + s_1 n + s_0,\ s_2 > 0; \textit{ and}$$

$$E(f) \geqslant \frac{1}{c(f) + \eta((f')^{1/2})}, \quad \textit{for a constant } \eta > 0, \textit{ for } c(f') \textit{ large.}$$

We turn to more general families of multi-point iterations. Let φ be a Hermite interpolatory iteration with $v(\varphi) = n$. Then $p(\varphi) \leqslant 2^{n-1}$ (Kung and Traub [5, Corollary 7.1]). Clearly, $a(\varphi) \geqslant n - 1$. Hence, by (3.9),

$$e(\varphi, f) \leqslant \frac{n-1}{nc_f + n - 1} \leqslant \frac{1}{c_f + 1}. \tag{5.7}$$

Since Ψ_n and ω_n are Hermite interpolatory iterations, from (5.7) and Corollaries 5.1, 5.2, we have

THEOREM 5.5. *For the family* Φ *of Hermite interpolatory iterations,*

$$\max\left(\frac{n-1}{nc(f) + r(n)}, \frac{n-1}{(n-1)c(f) + c(f') + s(n)}\right) \leqslant E_n(f) \leqslant \frac{n-1}{nc_f + n - 1}, \quad \forall n,$$

$$\max\left(\frac{1}{c(f)}\left[1 + \frac{\zeta}{(c(f))^{1/2}}\right], \frac{1}{c(f) + \eta(c(f'))^{1/2}}\right) \leqslant E(f) \leqslant \frac{1}{c_f + 1},$$

for c_f *large, where* $r(n) = r_2 n^2 + r_1 n + r_0, r_2 > 0, s(n) = s_2 n^2 + s_1 n + s_0,$ $s_2 > 0, \zeta < 0$ *and* $\eta > 0$.

REMARK 5.1. The lower and upper bounds for $E_n(f)$ and $E(f)$ stated in Theorem 5.5 are tight for f such that $c(f) \sim c_f$ and c_f is large, since lower bound/upper bound $\longrightarrow 1$ as $c_f \longrightarrow \infty$. In this case, by Theorem 5.2, Ψ_M is close to optimal among all Hermite interpolatory iterations.

Now, let φ be any multi-point iteration which uses evaluations of f only. Let $v(\varphi) = n$. Then $p(\varphi) \leqslant 2^n$ (Kung and Traub [5, Theorem 7.2]). Clearly, $a(\varphi) \geqslant n - 1$. Hence

$$e(\varphi, f) \leqslant \frac{n}{nc(f) + n - 1} \leqslant \frac{1}{c(f)}. \tag{5.8}$$

Since Ψ_n is a multi-point iteration which uses evaluations of f only, from (5.8) and Corollary 5.1, we have

THEOREM 5.6. *For the family Φ of multi-point iterations using values of f only,*

$$\frac{n-1}{nc(f) + r(n)} \leqslant E_n(f) \leqslant \frac{n}{nc(f) + n - 1}, \ \forall n,$$

$$\frac{1}{c(f)}\left[1 + \frac{\zeta}{(c(f))^{1/2}}\right] \leqslant E(f) \leqslant \frac{1}{c(f)},$$

for $c(f)$ large, where $r(n) = r_2 n^2 + r_1 n + r_0, r_2 > 0$, and $\zeta < 0$.

REMARK 5.2. The lower and upper bounds for $E_n(f)$ and $E(f)$ stated in Theorem 5.6 are tight for f such that $c(f)$ is large, since lower bound/upper bound $\longrightarrow 1$ as $c(f) \longrightarrow \infty$. In this case, by Theorem 5.2, Ψ_M is close to optimal among all multi-point iterations using values of f only.

REMARK 5.3. For a given problem f let $E'(f), E''(f)$ be the optimal efficiency achievable by one-point iteration and multi-point iteration, respectively. By Theorem 4.2 and Corollary 5.1,

$$E'(f) \leqslant \frac{\log 3}{3c_f + 2},$$

$$E''(f) \geqslant \frac{1}{c(f)}\left[1 + \frac{\zeta}{(c(f))^{1/2}}\right], \quad \zeta < 0, \text{ for } c(f) \text{ large.}$$

Hence

$$\frac{E''(f)}{E'(f)} \geqslant \frac{3c_f + 2}{(\log 3)c(f)}\left[1 + \frac{\zeta}{(c(f))^{1/2}}\right] \sim \frac{3}{\log 3} \cdot \frac{c_f}{c(f)}, \quad \text{for } c(f) \text{ large.}$$

In particular, if f is a problem such that $c_f = c(f)$ and c_f is large, then the ratio between optimal efficiencies achievable by multi-point iteration and one-point iteration is at least $3/\log 3 \sim 1.89$.

6. A conjecture. Kung and Traub [5] conjecture that if φ is any multi-point iteration with $v(\varphi) = n$ then $p(\varphi) \leq 2^{n-1}$. Suppose that this conjecture is true. Then, by (3.9), for any multi-point iteration φ with $v(\varphi) = n$,

$$e(\varphi, f) \leq (n-1)/(nc_f + a(\varphi)).$$

Clearly, $a(\varphi) \geq n - 1$. Hence

$$e(\varphi, f) \leq (n-1)/(nc_f + n - 1) \equiv k(n).$$

Observe that

$$k(n) \leq 1/(c_f + 1), \quad \forall n, \forall c_f.$$

Therefore we propose the following conjecture. *It states, essentially, that the optimal efficiency for solving the problem f with respect to all one-point and multi-point iterations is bounded by the reciprocal of the problem complexity.*

CONJECTURE 6.1. *For the family Φ of one-point or multi-point iterations,*

$$E_n(f) \leq (n-1)/(nc_f + n - 1), \quad E(f) \leq 1/(c_f + 1).$$

7. Numerical example. Let $f(x) = \Sigma_{i=1}^{50} ix^i - 25$. We calculate its simple zero $\alpha = -1$. Calculations were done in double precision arithmetic on a DEC PDP-10 computer. About 16 digits are available in double precision. Numerical results show the following: Starting with $x_0 = -1.01$, to bring the error to about 10^{-16}, five Newton-Raphson iterations are required while one ω_6 iteration is required. (See Table 7.1.) We assume that we do not take advantage of the algebraic dependence of f and f' (see the assumption of (3.5)) and that we use Horner's rule for the evaluation of f and f', treating each as an independent polynomial. Suppose that we use the procedure given by Kung and Traub [5, Appendix] to compute $\omega_6(f)(x)$.

Let Σ^1 and Σ^2 be algorithms associated to Newton-Raphson iteration and ω_6 respectively. Then the total costs are

$$k(\Sigma^1) = 5[2 \cdot 50 + 2 \cdot 49 + 2] = 10^3,$$

$$k(\Sigma^2) = 5 \cdot 2 \cdot 50 + 2 \cdot 49 + \frac{3}{2} \cdot 6^2 - 4 = 657;$$

and the efficiencies are

$$e(\Sigma^1) = 1/[2 \cdot 50 + 2 \cdot 49 + 2] = 5/10^3,$$

$$e(\Sigma^2) = 5/\left[5 \cdot 2 \cdot 50 + 2 \cdot 49 + \frac{3}{2} \cdot 6^2 + \frac{3}{2} \cdot 6 - 4\right] = 5/657.$$

Then

$$\frac{k(\Sigma^1)}{k(\Sigma^2)} = \frac{10^3}{657}, \quad \frac{e(\Sigma^2)}{e(\Sigma^1)} = \frac{10^3}{657},$$

as predicted by (2.2). (In general, *approximate* equality holds from (2.2).)

Let $x_{i+1} = \varphi(x_i)$. The errors when φ is Newton-Raphson and $\varphi = \omega_6$ are shown in Table 7.1.

	Newton-Raphson	ω_6
$x_0 - \alpha$	-1.0×10^{-2}	-1.0×10^{-2}
$x_1 - \alpha$	-2.1×10^{-3}	-2.2×10^{-16}
$x_2 - \alpha$	-1.0×10^{-4}	
$x_3 - \alpha$	-2.7×10^{-7}	
$x_4 - \alpha$	-1.8×10^{-12}	
$x_5 - \alpha$	-1.1×10^{-16}	

TABLE 7.1

Acknowledgement. We want to thank G. W. Stewart for his comments on the work reported in this paper.

Bibliography

1. W. M. Gentleman, Private communication, 1970.

2. D. Knuth, *The art of computer programming.* Vol. 2. *Seminumerical algorithms,* Addison-Wesley, Reading, Mass., 1969. MR 44 #3531.

3. H. T. Kung, *A bound on the multiplicative efficiency of iteration,* J. Comput. System Sci. 7 (1973), 334–342.

4. ———, *The computational complexity of algebraic numbers,* Proc. Fifth Annual ACM Sympos. on Theory of Computing, 1973, pp. 152–159; SIAM J. Numer. Anal. (to appear).

5. H. T. Kung and J. F. Traub, *Optimal order of one-point and multi-point iteration,* Department of Computer Science report, Carnegie-Mellon University, 1973; J. Assoc. Comput. Mach. (to appear).

6. A. M. Ostrowski, *Solution of equations and systems of equations,* 2nd ed., Academic Press, New York, 1966. MR 35 #7575.

7. M. S. Paterson, *Efficient iterations for algebraic numbers,* Complexity of Computer Computations, R. Miller and J. W. Thatcher (eds.), Plenum Press, New York, 1972, pp. 41–52.

8. J. F. Traub, *Iterative methods for the solution of equations,* Prentice-Hall, Englewood Cliffs, N. J., 1964. MR 29 #6607.

9. ———, *Computational complexity of iterative processes,* SIAM J. Comput. 1 (1972), 167–179. MR 47 #2860.

CARNEGIE-MELLON UNIVERSITY

Author Index

Roman numbers refer to pages on which a reference is made to an author or work of an author.

Italic numbers refer to pages on which a complete reference to a work by the author is given.

Boldface numbers indicate the first page of the articles in the book.

Subject Index